Series in BioEngineering

The Series in Bioengineering serves as an information source for a professional audience in science and technology as well as for advanced students. It covers all applications of the physical sciences and technology to medicine and the life sciences. Its scope ranges from bioengineering, biomedical and clinical engineering to biophysics, biomechanics, biomaterials, and bioinformatics.

Indexed by WTI Frankfurt eG, zbMATH.

Khin Wee Lai · Pauline Shan Qing Yeoh
Editors

Biomedical Engineering

AI and Technological Innovations

Springer

Editors
Khin Wee Lai
Department of Biomedical Engineering
Universiti Malaya
Kuala Lumpur, Malaysia

Pauline Shan Qing Yeoh
Department of Biomedical Engineering
Universiti Malaya
Kuala Lumpur, Malaysia

ISSN 2196-8861 ISSN 2196-887X (electronic)
Series in BioEngineering
ISBN 978-981-97-9296-2 ISBN 978-981-97-9294-8 (eBook)
https://doi.org/10.1007/978-981-97-9294-8

This Springer imprint is published by the registered company Springer Nature Singapore Pte Ltd.
The registered company address is: 152 Beach Road, #21-01/04 Gateway East, Singapore 189721, Singapore

Preface

The healthcare industry is experiencing a significant shift due to progress in artificial intelligence (AI) and new technologies. These innovations are revolutionizing how medical professionals diagnose, treat, and monitor patients, leading to improved outcomes and more efficient care. This book has two primary aims: first, it delivers an in-depth analysis of critical areas where AI and emerging technologies are making notable advancements in healthcare; second, it explores the application of these technologies in various healthcare contexts, including maternal care, prosthetics, and chronic disease management. By covering both theoretical research and practical applications, the book provides a thorough perspective on the rapid developments at the intersection of AI, technology, and healthcare, effectively bridging the gap between research and real-world practice.

The book begins by exploring the impact of emerging technologies on maternal and neonatal care. The chapter "Development of a Web-Based Maternal Health Monitoring System: Advancing Beyond Manual Record-Keeping" examines how digitizing electronic medical records can improve maternal health outcomes by integrating these records into web-based applications. This shift from traditional manual record-keeping to digital solutions represents a broader trend toward healthcare digitization, a theme that recurs throughout the book. The chapter emphasizes the importance of user-friendly platforms for both patients and healthcare providers, particularly addressing the specific needs of maternal healthcare. It ensures that all critical data is securely stored and easily accessible, thereby enhancing the overall quality of care and communication within the healthcare ecosystem. In the context of patient care, the chapter "Development and Testing of a Home-Based Phototherapy System for Neonatal Jaundice" highlights how technology can improve access to care through the development of a home-based phototherapy system with an automated irradiance light mechanism based on clinical guidelines. This system allows parents to manage neonatal jaundice at home, reducing hospital visits and costs. This innovation showcases how technology can empower patients and enhance the quality of care for common but treatable conditions.

The chapter "Non-invasive Blood Glucose Monitoring via Infrared Absorbance" delves into emerging technologies for managing chronic diseases. It validates the use of near-infrared absorption as a method for non-invasive glucose monitoring. This study explores the feasibility of a simple prototype model using widely available infrared emitters, such as 940 nm LEDs, and laboratory samples of distilled water mixed with dextrose, which closely resembles glucose in blood. The investigation focuses on assessing the sensitivity and selectivity of this technology and examines basic optimization techniques to understand how minor adjustments in calibration can affect measurement accuracy. These advancements highlight how non-invasive glucose monitoring can benefit individuals with diabetes and those requiring frequent glucose assessments in clinical settings.

AI's integration into healthcare spans numerous applications, including machine learning models for diagnostics, smart wearables, robotic surgeries, and personalized medicine. A crucial aspect of AI's impact is its ability to prevent failures in medical equipment. The chapter "Empowering Predictive Maintenance of Medical Equipment Through AI-Driven Condition Monitoring" investigates how AI can monitor equipment performance and predict potential malfunctions before they occur. This capability helps ensure that essential medical devices remain functional and reliable, extends the lifespan of critical medical devices, and optimizes healthcare delivery.

Medical imaging is another key focus of the book, with several chapters addressing AI's influence in this field. The chapter "Preliminary Analysis of Ultrasound Features for Detection of Polycystic Ovary Syndrome (PCOS) in Women" explores the challenges of diagnosing PCOS by investigating various feature extraction techniques for its detection through ultrasound imaging. The chapter "Multi-label Convolutional Neural Network for Multiple Diseases Detection in Chest Radiographs" demonstrates how AI can analyze medical images more quickly and accurately than traditional methods, transforming diagnostic radiology. Another chapter "Application of Model-Agnostic Meta Learning (MAML) in Automated Invasive Ductal Carcinoma Grading", explores how few-shot learning can enhance cancer detection, enabling more accurate diagnoses even with limited datasets.

In prosthetics, AI is playing a transformative role, as discussed in chapter "Artificial Intelligence in Prosthetic Gait Assessment and Prosthesis Control: Present and Future". AI is improving the design and control of prosthetic devices, particularly for lower-limb prosthetics. This chapter reviews the application of AI-based methods in assessing the gait of lower-limb amputees. It also highlights the use of AI models in advanced control systems for powered prostheses and discusses future prospects for smart prostheses. The book concludes with discussions on unconventional therapies, with the chapter "Mechanomyography Analysis of Respiratory Muscle Function While Singing in Spinal Cord Injury" presenting an innovative approach to rehabilitation, utilizing singing to train respiratory muscles.

In summary, this book, *Biomedical Engineering: AI and Technological Innovations*, part of the *Series in BioEngineering*, provides a comprehensive overview of recent advancements, designed to inspire further exploration and innovation. Whether you are a researcher, clinician, or student, this book offers a deep understanding of how AI and emerging technologies are shaping the future of healthcare and provides valuable insights into the possibilities and challenges ahead.

Kuala Lumpur, Malaysia

Khin Wee Lai
Pauline Shan Qing Yeoh

Contents

Development of a Web-Based Maternal Health Monitoring System: Advancing Beyond Manual Record-Keeping

Azira Khalil, Ahmad Luqmanulhakim Ahmad Sunawari, Shahrina Ismail, and Aminatul Saadiah Abdul Jamil

Abstract The Web-Based Maternal Health Monitoring System represents a significant advancement in the digitalisation of maternal health records, addressing critical limitations in Malaysia's traditional paper-based system. The existing manual record-keeping process is not only time-consuming but also prone to wear and data loss, and it lacks the accessibility required for seamless healthcare across different facilities. E-Maternity provides a web-based platform that allows pregnant women to easily track their maternal health records and progress via mobile devices, improving the continuity of care and reducing the administrative burden on healthcare providers. The system's design emphasises user-friendly access to comprehensive maternal health information, yet it faces several challenges. The reliance on internet connectivity poses a barrier, particularly in rural areas with poor network coverage, and the absence of two-factor authentication raises concerns about data security. Additionally, the system's dependency on local storage limits its scalability and efficiency. This chapter explores the development, implementation, and testing of the E-Maternity system, highlighting both its strengths and limitations. Functional testing was conducted to ensure the system met key performance criteria, with a focus on usability, reliability, and security. The results demonstrated the system's potential to enhance maternal healthcare delivery, although certain limitations were identified. Future recommendations include expanding the system's functionality to cover postnatal care, integrating cloud-based storage to enhance data management and security, and implementing two-factor authentication to strengthen user data protection. By addressing these challenges, the E-Maternity system can evolve into a more robust and secure platform, offering a comprehensive solution for maternal health monitoring in Malaysia. This study contributes to the ongoing efforts to improve healthcare delivery through digital innovation, particularly in regions where traditional methods are insufficient.

Keywords E-Maternity system · Electronic Medical Records (EMR) · Maternal health records · Web-Based application · Functional testing

---

A. Khalil (✉) · A. L. Ahmad Sunawari · S. Ismail · A. S. Abdul Jamil
Faculty of Science and Technology, Universiti Sains Islam Malaysia, Bandar Baru Nilai, 71800 Nilai, Negeri Sembilan, Malaysia
e-mail: azira@usim.edu.my

K. W. Lai and P. S. Q. Yeoh (eds.), *Biomedical Engineering*, Series in BioEngineering,
https://doi.org/10.1007/978-981-97-9294-8_1

1 Introduction

Imagine a world where crucial health data is at risk of being lost, misplaced, or damaged, all because it's stored in a fragile, outdated logbook. In Malaysia, pregnant mothers are still dependent on these physical records to document their pregnancy journey, carrying them from one clinic to another [1]. With each visit, another handwritten entry is added to a paper-based system that has not evolved in decades. But what happens when that logbook is forgotten, damaged, or lost? The consequences could be severe, impacting the timely care of both mother and child. In an era where digital solutions are transforming every aspect of our lives, why should maternal health care be left behind? This chapter addresses the critical need for a digital revolution in maternal health records, ensuring that vital information is secure, accessible, and always at the fingertips of those who need it most.

In Malaysia, paper-based medical records have traditionally been the primary method for recording patient data, including at the Maternal and Child Health Clinic [2]. Pregnancy is a critical period for both mother and baby, necessitating frequent monitoring and care. The typical length of a pregnancy is approximately 280 days, or 40 weeks, from the last menstrual period [3]. Malaysian mothers are required to report their pregnancies at a minimum of 10 weeks to the nearest clinic, where healthcare providers monitor both maternal and fetal health [4]. Nurses are responsible for scheduling regular check-ups and documenting all collected data in a designated logbook known as the Maternal Health Record [5].

In 2020, 470,195 babies were born in Malaysia, correlating to an equivalent number of Maternal Health Records (Department of Statistics Malaysia, 2021). These records, vital for tracking pregnancy progression, are kept by both the mother and the healthcare facility. Mothers are required to bring these records to all prenatal appointments and when traveling, to ensure that critical health information is available in case of an emergency. However, this manual system has notable drawbacks, particularly in the digital age. The A5-sized paper logbooks are cumbersome, prone to wear and tear, and easily misplaced. Furthermore, the lack of synchronisation between hospitals and clinics can lead to fragmented care and potential data loss if the logbook is unavailable [6].

Given the growing emphasis on digitalisation in healthcare, Electronic Medical Records (EMR) present a compelling alternative for managing pregnancy records in Malaysia. The primary goal of EMR systems is to offer a more organised and efficient method for maintaining patient data [7]. EMRs enable healthcare providers to share vital health information, ensuring coordinated and comprehensive care for pregnant women. These systems, often cloud-based, facilitate data accessibility across multiple healthcare facilities [8]. However, traditional EMRs are typically hospital-based and inaccessible to patients, limiting their ability to track their pregnancy journey [9].

To address these limitations, integrating EMR into a web-based application offers a practical solution. Such an application can digitalise all maternal health records, making them accessible to both healthcare providers and mothers, thus ensuring continuous and coordinated care throughout the pregnancy. This approach eliminates the need for physical logbooks, reduces the risk of data loss, and enhances convenience for mothers and healthcare providers alike.

2 Electronic Medical Records (EMR)

The evolution of healthcare documentation has seen a significant shift from traditional paper-based records to digital solutions, with Electronic Medical Records (EMR) at the forefront of this transformation. EMRs are digital versions of patients' paper charts, encompassing a comprehensive repository of medical histories, diagnoses, medications, treatment plans, immunisation dates, allergies, radiology images, and laboratory test results [10]. Beyond mere storage, EMRs facilitate streamlined access to patient data, enhancing the efficiency and quality of healthcare delivery.

In Malaysia, the journey towards EMR adoption began in earnest during the Sixth Malaysian Plan with the establishment of the EMR system in 1993. This initiative culminated in the implementation of EMRs at Selayang Hospital in 1999, encompassing clinical, imaging, and administrative functionalities. The subsequent Seventh Malaysian Plan saw the expansion of this digital framework, integrating Hospital Information Systems (HIS) as foundational components [11].

Recognising the transformative potential of EMRs, the Malaysian government has articulated ambitious goals within the Twelfth Malaysia Plan (RMK12), aiming to digitalise citizens' lifetime health records and ensure their accessibility across a network of health facilities [12]. This vision aligns with global trends emphasising interoperability, patient-centered care, and data-driven decision-making in healthcare.

Despite these aspirations, the path to comprehensive EMR integration has encountered challenges. As of July 2020, only 37 out of 146 government hospitals (approximately 25%) and 97 out of 1,090 public health clinics (around 9%) had adopted EMR systems [12]. These figures underscore the need for accelerated efforts, infrastructural investments, and stakeholder engagement to realise the full benefits of EMRs.

The advantages of EMRs are manifold:

1. Improved Accessibility: EMRs ensure that patient data is readily available to authorised healthcare providers, facilitating timely interventions and reducing redundancies [13].
2. Enhanced Coordination: Digital records promote seamless communication among multidisciplinary teams, ensuring that all caregivers are informed and aligned in their treatment approaches [14].
3. Data Security and Integrity: EMRs incorporate security protocols, including encryption and access controls, safeguarding sensitive patient information against unauthorised access and data breaches [15].

4. Analytical Capabilities: The digitisation of health records enables advanced data analytics, supporting population health management, research initiatives, and policy formulation [16].
5. Patient Empowerment: Modern EMR systems often include patient portals, granting individuals access to their health information, fostering engagement, and promoting informed decision-making [17].

In the context of maternal health, integrating EMRs presents a transformative opportunity. By replacing the traditional Maternal Health Record logbooks with digital counterparts, Malaysia can address issues of data fragmentation, accessibility challenges, and the risks associated with physical record-keeping. Moreover, leveraging cloud-based EMR solutions can ensure that maternal health data is synchronised across clinics and hospitals, enabling continuous and coordinated care for expectant mothers.

As Malaysia progresses towards a digitally empowered healthcare ecosystem, the comprehensive adoption of EMRs stands as a pivotal milestone, promising enhanced care quality, operational efficiency, and improved health outcomes.

3 System Design and Implementation of E-Maternity System as the Web-Based Maternal Health Monitoring System

The E-Maternity system is a web-based application designed to streamline maternal health monitoring by replacing traditional manual record-keeping with an automated, user-friendly platform. The system facilitates seamless interaction between patients, healthcare providers, and administrators, allowing for efficient management of medical records, appointments, and patient information. It is built to cater to the specific needs of maternal healthcare, ensuring that all critical data is securely stored and easily accessible, thereby enhancing the overall quality of care and communication within the healthcare ecosystem.

3.1 Waterfall Model Methodology

The development of the E-Maternity web-based system followed a structured approach using the Software Development Life Cycle (SDLC) model, specifically the Waterfall Model. Introduced in 1970, the Waterfall Model emphasises a linear and sequential design process, which is ideal for projects with well-defined requirements. This method comprises five distinct stages: requirement analysis, system design, implementation, testing, and maintenance. Each phase must be completed before the next one begins, ensuring that the project progresses in an orderly fashion,

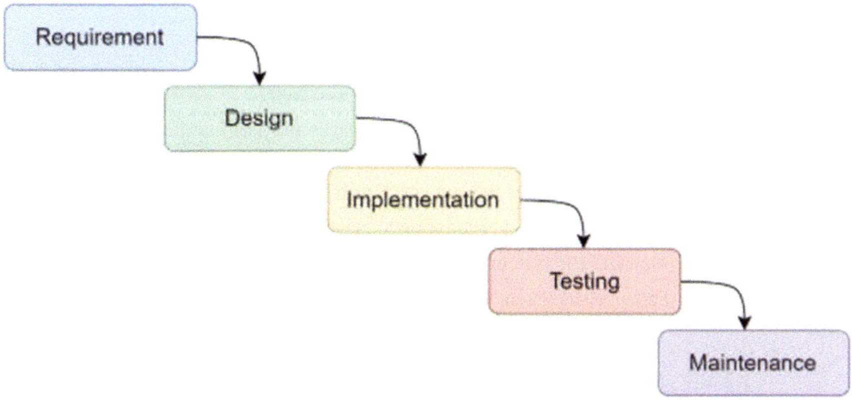

Fig. 1 Waterfall model of software development life cycle (SDLC)

reducing the risk of overlooking critical details. Figure 1 illustrates the Waterfall Model's stages as applied to the E-Maternity system.

3.1.1 Requirement Analysis

The requirement analysis phase is fundamental to the success of the project, as it defines what the system should achieve and the specific features it must include. In this phase, the focus is on understanding the problem the software intends to solve and outlining the functional requirements that will address this problem [18]. For the E-Maternity system, the analysis involved gathering detailed information through online research to define the functionality that the system would provide, ensuring alignment with the needs of the end-users.

3.1.2 Design

The design phase translates the requirements into a blueprint for the system. This includes creating detailed specifications for the software's architecture, encompassing both functional and non-functional requirements. For the E-Maternity system, the Unified Modelling Language (UML) was employed to create visual models, such as data flow diagrams and use case diagrams, which help communicate the system's design clearly to stakeholders. The design document produced in this phase serves as a guide for the subsequent implementation phase.

Table 1 Programming language to use in developing E-Maternity

Language	Functions
HTML	Structures the content and layout of web pages
JavaScript	Enhances the visual appeal of web pages and user interfaces by adding browser-specific HTML or CSS styles
CSS	Designs and styles the content of web pages
PHP	Processes server-side scripting to generate dynamic HTML output
phpMyAdmin	Manages the system's database
Apache	Serves web content, delivering requested images, files, and other data to users

3.1.3 Implementation

Implementation is the phase where the design is translated into a working system. The E-Maternity system was developed using a combination of HTML, CSS, PHP, JavaScript, and phpMyAdmin, each chosen for its specific role in creating a robust web-based solution. HTML and CSS were used for structuring and styling the web pages, while PHP and JavaScript handled the dynamic aspects of the application. phpMyAdmin served as the database management tool. Table 1 summarises the programming languages and tools utilised during this phase.

3.1.4 Testing

Testing is crucial to ensure that the E-Maternity system meets its specified requirements and is free of defects. This phase includes several testing types, such as unit testing, integration testing, system testing, and acceptance testing, each designed to validate different aspects of the system. The goal is to deliver a reliable and functional system ready for deployment.

3.1.5 Maintenance

The final phase of the Waterfall Model is maintenance, which involves ongoing support for the system after deployment. The software will be evaluated based on its requirements to confirm that it functions as intended per the specifications [19]. This includes fixing any bugs, updating the software, and optimising performance. Additionally, as user needs evolve, new features or enhancements may be added to the E-Maternity system during this phase.

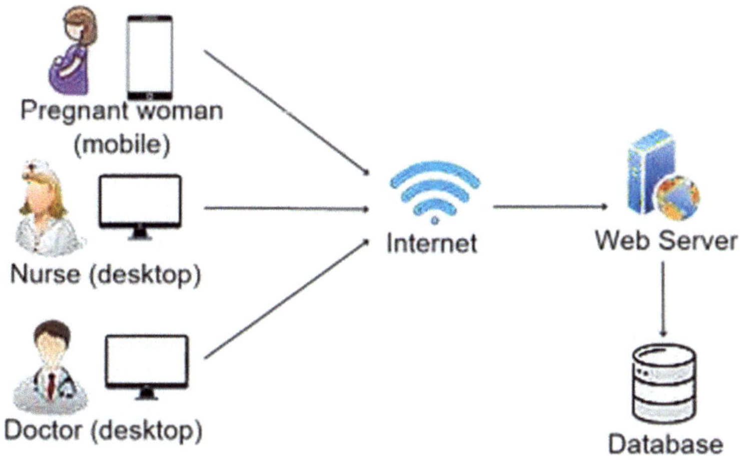

Fig. 2 System architecture of E-Maternity

3.2 *System Architecture*

The system architecture of the E-Maternity application defines how various components interact within the system. Figure 2 illustrates the architecture, showing the interaction between users (patients, doctors, nurses, and admin) and the system through the user interface. The application relies on a web server to host the system and retrieve data from the database server, which is managed using phpMyAdmin.

3.3 *Use Case Diagram*

The use case diagram (Fig. 3) provides a high-level overview of the interactions between the users and the E-Maternity system. It captures the functional requirements by illustrating the roles of doctors, nurses, and patients within the system. The admin is responsible for managing user accounts, while doctors and nurses can update patient information and create new patient accounts. Patients can only view their medical records, ensuring data integrity.

3.4 *Flowchart Diagram*

Flowcharts are used to visualise the process flow within the E-Maternity system, detailing the actions and decisions that users go through. Separate flowcharts are created for patients, doctors, and nurses to reflect their specific interactions with the

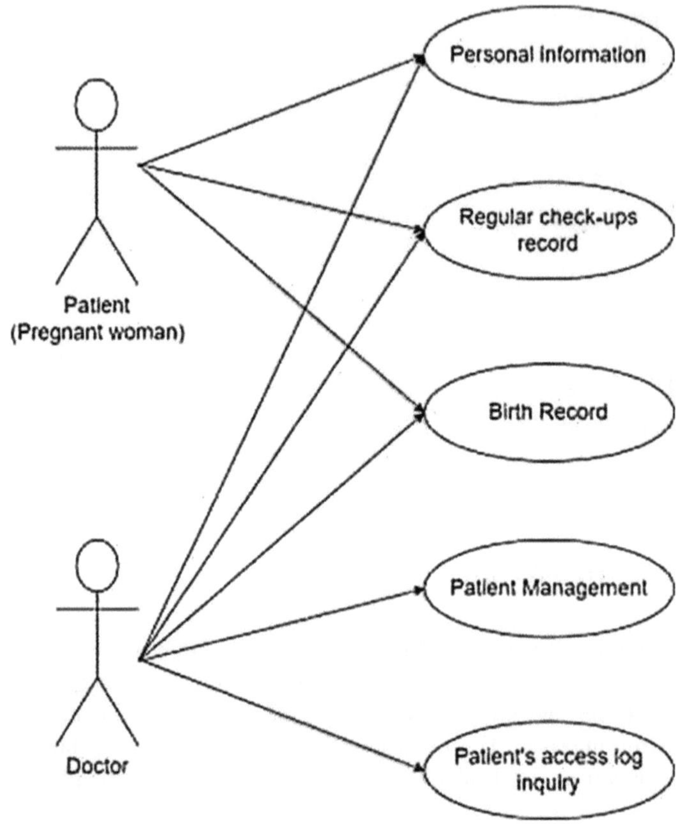

Fig. 3 Use case diagram for E-Maternity

system. Figure 4. illustrates the flowchart for doctors and nurses, while Fig. 5 shows the patient's flowchart.

4 Comprehensive Analysis and Evaluation of the E-Maternity System

This section delves into the various aspects of the E-Maternity system, covering everything from its database architecture to user interface designs, functional testing, and feedback mechanisms. The discussion aims to provide an in-depth understanding of the system's structure, functionality, user interaction, and overall performance. By exploring these facets, we can assess the system's effectiveness in improving maternal health monitoring and identify areas that might require further refinement.

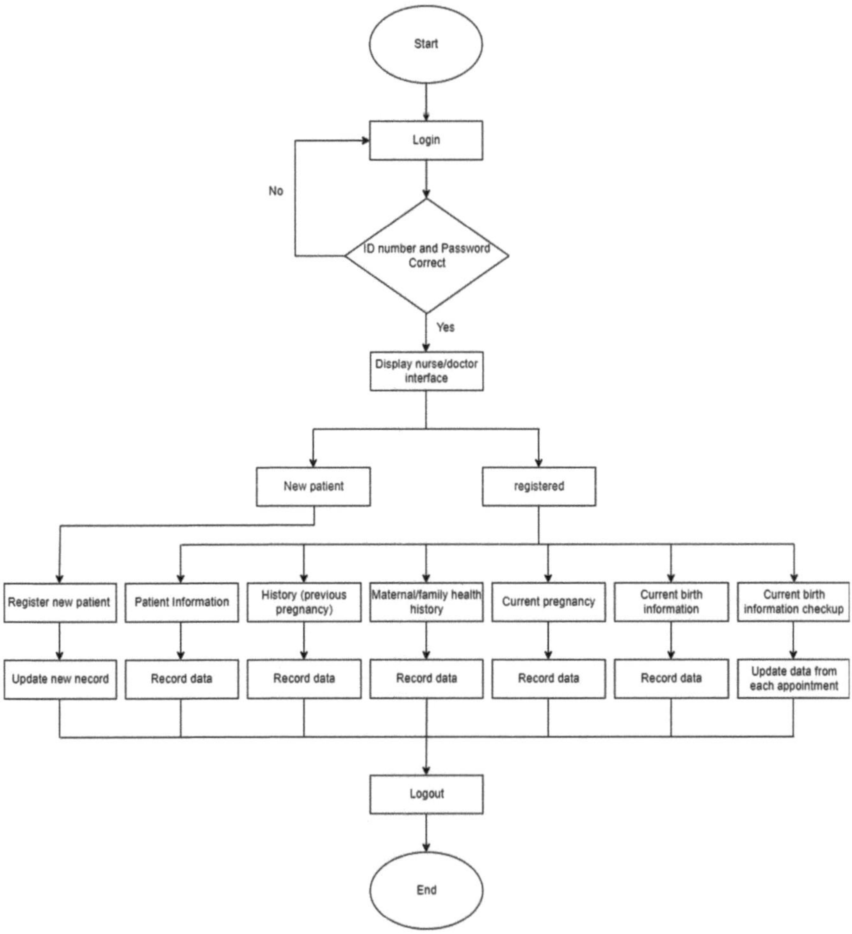

Fig. 4 Flowchart of E-Maternity system for doctor and nurse

4.1 E-Maternity System Database

The database is the backbone of the E-Maternity system, responsible for storing and managing all data related to maternal health records, user accounts, and system activities. MySQL was chosen as the database management system due to its reliability, scalability, and wide adoption in web-based applications. The database is organised into various tables that store patient information, health records, user roles, and system logs, among other data.

One of the key strengths of the E-Maternity system's database design is its ability to handle large volumes of data while maintaining quick retrieval times. The relational nature of MySQL ensures that data is stored in an organised manner, with relationships defined between different tables. This setup allows for complex queries and

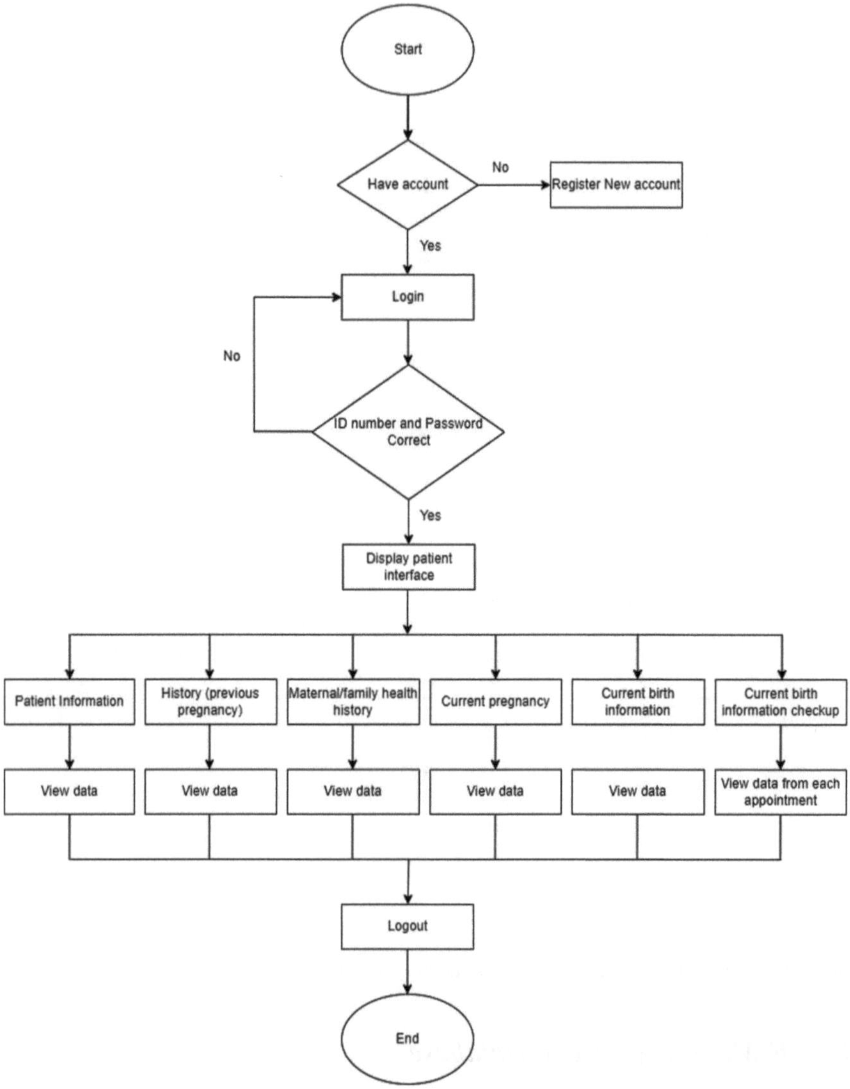

Fig. 5 Flowchart of E-Maternity system for patient

data retrieval processes, which are essential for generating comprehensive maternal health reports and tracking patient progress over time.

Moreover, the system includes mechanisms for data integrity and security. Data validation processes ensure that only accurate and complete data is entered into the system, reducing the risk of errors. Additionally, the use of secure connections between the web application and the database helps protect sensitive patient information from unauthorised access.

Data Storage Requirement:

$$S_{total} = N_{users} x \, D_{record}$$

where:

S_{total} is the total storage required.
N_{users} is the number of users (e.g., pregnant women).
D_{record} is the average size of one user's digital maternal health record.

4.2 *Module Design*

The E-Maternity system is designed to cater to different types of users, each with specific roles and access levels. These roles include administrators, doctors, nurses, and patients. The system's user interface is tailored to meet the needs of each user group, ensuring that the system is both functional and user-friendly.

4.2.1 Admin Interface

Administrators play a crucial role in managing the overall system. Their responsibilities include creating and managing user accounts for hospitals, doctors, nurses, and patients. The admin interface is designed to be intuitive, allowing administrators to easily navigate through the system's features. This interface provides a comprehensive overview of all registered users and allows administrators to perform essential tasks such as adding new accounts, updating existing user information, and deleting accounts that are no longer in use.

4.2.2 Doctor and Nurse Interface

Doctors and nurses are the primary users of the E-Maternity system, as they are directly involved in patient care. The interface for these users is designed to provide quick access to patient information and maternal health records. This includes viewing a patient's medical history, updating health records after consultations, and registering new patients into the system. The design prioritises ease of use, ensuring that healthcare providers can efficiently document and retrieve patient information without unnecessary delays.

4.2.3 Patient Interface

The patient interface is designed to empower patients by providing them with access to their own maternal health records. Patients can view their health records, track updates following medical consultations, and monitor the progress of their pregnancy. However, to maintain the integrity of the medical records, patients are not allowed to modify any health data. The interface is designed to be simple and easy to navigate, even for users who may not be technologically proficient.

4.3 System Feedback

Feedback mechanisms are an essential component of the E-Maternity system, as they guide users through the system and help prevent errors. Feedback is provided in two main forms: status information and warning messages.

4.3.1 Status Information

Status information is provided to users after they complete an action, such as registering a new patient or updating a health record. This feedback serves as confirmation that the action has been successfully completed, ensuring that users are aware of the outcomes of their interactions with the system. By providing immediate feedback, the system helps build user confidence and reduces the likelihood of repeated actions or errors.

4.3.2 Warning Messages

Warning messages are designed to prevent errors by alerting users when they fail to complete a required action or enter incorrect data. For example, suppose a user attempts to submit a form without filling in all the necessary fields. In that case, a warning message will be displayed, prompting the user to correct the issue before proceeding. These messages are crucial for maintaining data accuracy and ensuring that the system functions as intended.

4.4 Functional Testing

Functional testing is crucial for validating that the E-Maternity system meets its specified requirements and operates as intended. This testing involves evaluating the system's components and functionalities against predefined criteria to ensure they perform correctly. The testing is conducted by creating detailed test cases that compare the expected results with actual system behavior.

Table 2 provides a comprehensive checklist for functional testing, detailing various test cases and their outcomes. This checklist covers a range of functionalities within the E-Maternity system, including user login, patient registration, data management, and error handling. Each test case is designed to verify specific features and ensure that the system behaves as expected under various conditions.

Test Case Overview.

- Admin Login: Tests the login functionality for administrators using valid and invalid credentials to confirm that access control is correctly implemented.
- Add New Doctor/Nurse: Verifies that new doctor and nurse entries can be successfully added to the system and appear in the list of users.

Table 2 Functional testing checklist

Test Case ID	Function	Description	Expected Result
TC-01	Admin login	Verify login functionality with correct credentials	Successful login with dashboard access
TC-02	Admin login	Verify login with incorrect username	Error message: "Invalid username or password."
TC-03	Admin login	Verify login with incorrect password	Error message: "Invalid username or password."
TC-04	Add new doctor/nurse	Add a new doctor/nurse to the system	New doctor/nurse appears in the list
TC-05	Register patient	Register a new patient	New patient is listed with assigned ID
TC-06	Update patient record	Update patient's pregnancy information	Changes are saved and reflected in the patient's record
TC-07	View patient record	View patient's health records	Patient records are displayed accurately
TC-08	Delete doctor/nurse	Delete a doctor/nurse from the system	Doctor/nurse is removed from the list
TC-09	Password change	Change user's password	Password is updated successfully
TC-10	Access control	Verify user role access (admin, doctor, nurse, patient)	Correct access level and permissions applied
TC-11	Form validation	Check form validation (required fields, correct format)	Appropriate error messages for invalid inputs
TC-12	Error handling	Handle system errors and exceptions	System displays user-friendly error messages
TC-13	System logout	Verify logout functionality	User is logged out and redirected to the login page
TC-14	Data integrity	Verify data consistency after operations (add, update, delete)	Data remains accurate and consistent across the system

- Register Patient: Ensures that new patients can be registered accurately, with their details correctly saved and displayed.
- Update Patient Record: Assesses whether patient records can be updated and whether changes are properly reflected in the system.
- View Patient Record: Checks that patient records are accessible and displayed accurately when requested.
- Delete Doctor/Nurse: Tests the deletion process to confirm that the removed doctor and nurse records are no longer listed in the system.
- Password Change: Validates the password change functionality for users, ensuring that new passwords are accepted and old ones are updated.
- Access Control: Evaluates role-based access to ensure that users can only access functionalities appropriate to their role (admin, doctor, nurse, patient).
- Form Validation: Checks for correct validation of forms, including mandatory fields and proper input formats, to prevent submission errors.
- Error Handling: Verifies that the system handles errors gracefully, providing user-friendly messages for incorrect actions or system issues.
- System Logout: Confirms that users can log out successfully and are redirected to the login page.
- Data Integrity: Ensures data consistency and accuracy after performing various operations such as adding, updating, and deleting records.

By systematically applying these test cases, the functional testing process helps to identify and address potential issues in the system, ensuring that the E-Maternity system operates reliably and effectively for its users.

To estimate the bandwidth needed to ensure smooth operation, especially in areas with limited internet connectivity, the following equation can be applied.

$$B_{required} = N_{concurrent} x \frac{D_{record}}{T_{access}}$$

where:

$B_{required}$ is the required bandwidth.
$N_{concurrent}$ is the number of concurrent users accessing the system.
D_{record} is the data size per access.
T_{access} is the average time taken to access a record.

4.5 User Acceptability Survey

To gauge the feasibility and usability of the E-Maternity system, a pilot user acceptability survey was conducted among a sample of healthcare providers and patients. The survey aimed to assess the overall user experience, including ease of use, system responsiveness, and satisfaction with the available features.

Participants in the survey provided valuable feedback on various aspects of the system. Healthcare providers, including doctors and nurses, were asked about the system's impact on their workflow and its effectiveness in improving patient care. Patients, on the other hand, were asked about their experience accessing and viewing their health records.

The survey results indicated a high level of acceptability among users, with most participants expressing satisfaction with the system's design and functionality. The positive feedback from both healthcare providers and patients suggests that the E-Maternity system is well-suited for its intended purpose. However, the survey also highlighted areas for potential enhancement, such as improving the user interface for less tech-savvy users and increasing the speed of certain operations.

5 Conclusion

In summary, the E-Maternity system presents a significant advancement over traditional paper-based maternal health record-keeping systems in Malaysia. Traditional systems are prone to wear and inefficiency due to manual data entry and limited accessibility across healthcare facilities. In contrast, E-Maternity offers a digital solution that enables pregnant women to track their maternal health records and progress conveniently via their mobile devices. This accessibility not only improves the continuity of care but also alleviates the burden on healthcare providers by reducing manual record-keeping and storage issues.

Despite its benefits, the E-Maternity system faces several limitations. Notably, the absence of two-factor authentication poses a security risk, as does the dependency on internet connectivity, which can be problematic in areas with poor network coverage. Additionally, the system's reliance on local storage can hinder its performance and accessibility, particularly for users with slow internet connections.

5.1 Future Implementation Recommendations

5.1.1 Larger Access

As the E-Maternity system expands to support a broader range of users and functionalities, including postnatal care and child health records, system performance becomes increasingly critical. One key aspect of performance is system latency, which refers to the delay experienced between a user action and the corresponding system response. High latency can significantly affect user experience, particularly in regions with slower internet connections or during peak usage times.

To better understand and address these challenges, it is essential to analyse the factors contributing to system latency. The total latency experienced by a user is influenced by several variables, including server response time, data transfer rate, and the size of the data being processed [20]. By examining these factors through a system latency equation, we can identify areas for optimisation that will enhance the overall responsiveness of the E-Maternity system.

$$L_{total} = L_{network} + L_{server} + L_{database}$$

where:

L_{total} is the total system latency.
$L_{network}$ is the latency due to network delay.
L_{server} is the latency caused by server processing time.
$L_{database}$ is the latency from database query processing.

5.1.2 Enhanced Security Measures

Implementing two-factor authentication (2FA) would significantly bolster the system's security. 2FA requires users to provide two forms of identification—such as a password and a verification code sent to a mobile device—before accessing their accounts. This additional layer of security helps to safeguard against unauthorised access and phishing attacks, ensuring that user data remains protected.

5.1.3 Cloud-Based Storage

Adopting cloud-based storage solutions could address limitations related to data storage and security. Cloud storage offers the advantage of remote data access, reducing the risk of data loss due to hardware failure or other physical issues. It also enhances data security through stringent authentication and access control measures, helping to prevent unauthorised access and data breaches. Moving to cloud-based storage would not only improve system reliability but also ensure that sensitive information is securely maintained and readily accessible.

By addressing these limitations and incorporating these recommendations, the E-Maternity system can evolve to better meet the needs of its users and provide a more robust, secure, and user-friendly maternal health monitoring solution.

Acknowledgements The authors would like to thank the Faculty of Science and Technology, Universiti Sains Islam Malaysia (USIM) for supporting the current study.

References

1. Norhayati, M.N., Nik Hazlina, N.H., Asrenee, A.R., Sulaiman, Z.: The experiences of women with maternal near miss and their perception of quality of care in Kelantan, Malaysia: a qualitative study. BMC Pregnancy Childbirth **17**, 1–14 (2017). https://doi.org/10.34172/jcs.2024.31796
2. Ameyaw, E.K., Amoah, P.A., Ezezika, O.: Effectiveness of mHealth apps for maternal health care delivery: systematic review of systematic reviews. J. Med. Internet Res. **26**, e49510 (2024). https://doi.org/10.2196/49510
3. Mangla, M., Anne, R.P.: Perinatal management of pregnancies with fetal congenital anomalies: a guide to obstetricians and pediatricians. Curr. Pediatr. Rev.. Pediatr. Rev. **20**(2), 150–165 (2024). https://doi.org/10.2174/1573396318666221005142001
4. Ministry of Health Malaysia.: Buku panduan penggunaan dan pengisian buku rekod kesihatan ibu hamil (4th ed.) (2020)
5. World Health Organization, & United Nations Children's Fund. Strengthening implementation of home-based records for maternal, newborn and child health: a guide for country programme managers. World Health Organ. (2023).
6. Srivastava, D., Mahanag, N.:. Blockchain in healthcare department: blockchain in healthcare data management. In Ubiquitous Comput. Technol. Innov. Univers. Healthc. IGI Global. pp. 274–304 (2024). https://doi.org/10.4018/979-8-3693-2268-0.ch013
7. Chiem, J.L., Hansen, E.E., Fernandez, N., Merguerian, P.A., Parikh, S.R., Reece, K., Low, D.K., Martin, L.D.: Transforming into a learning health system: a quality improvement initiative. Pediatr. Qual. & Saf. **9**(3), e724 (2024). https://doi.org/10.1097/pq9.0000000000000724
8. Chang, C.W., Ma, T.Y., Choi, M.S., Hsu, Y.Y., Tsai, Y.J., Hou, T.W.: Electronic personal maternity records: Both web and smartphone services. In Comput. Methods Programs Biomed. Elsevier Ireland Ltd. **121**(1), pp. 49–58. (2015) https://doi.org/10.1016/j.cmpb.2015.02.008
9. Putri, L.D., Girsang, E., Lister, I.N.E., Kung, H.T., Kadir, E.A., Rosa, S.L.: Public health implications for effective community interventions based on hospital patient data analysis using deep learning technology in indonesia. Information **15**(1), 41 (2024). https://doi.org/10.3390/info15010041
10. Fennelly, O., Moroney, D., Doyle, M., Eustace-Cook, J., Hughes, M.: Key interoperability Factors for patient portals and Electronic health Records: A scoping review. Int. J. Med. Inform., 105335. (2024). https://doi.org/10.1016/j.ijmedinf.2023.105335
11. Salleh, M.I.M., Abdullah, R., Zakaria, N.: Evaluating the effects of electronic health records system adoption on the performance of Malaysian health care providers. BMC Med. Inform. Decis. Mak.Decis. Mak. **21**(1), 1–13 (2021). https://doi.org/10.1186/s12911-021-01447-4
12. Economic Planning Unit.: Twelfth Malaysia Plan 2021–2025. (2021). https://rmke12.ekonomi.gov.my/en
13. Okolo, C.A., Ijeh, S., Arowoogun, J.O., Adeniyi, A.O., Omotayo, O.: Reviewing the impact of health information technology on healthcare management efficiency. Int. Med. Sci. Res. J. **4**(4), 420–440 (2024). https://doi.org/10.51594/imsrj.v4i4.1000
14. Nelson, L.H., Shue, J.B., Dash, K.:. Engaging key stakeholders in the development of a framework & toolkit for Age-Friendly Cross-Sector care coordination: three keys for aligning the health care & public health sectors. Soc. Innov. J., **24** (2024)
15. Akangbe, R., Charles-Chinkata, T.: Dealing with data breaches on patient's EMR sensitive data: a comprehensive approach. Front. Digit. Health (2024)
16. Adenyi, A.O., Okolo, C.A., Olorunsogo, T., Babawarun, O.: Leveraging big data and analytics for enhanced public health decision-making: A global review. GSC Adv. Res. Rev. **18**(2), 450–456 (2024). https://doi.org/10.30574/gscarr.2024.18.2.0078
17. Rahman, M.A., Jim, M.M.I.: Addressing privacy and ethical considerations In Health Information Management Systems (IMS). Int. J. Health Med. **1**(2), 1–13 (2024). https://doi.org/10.62304/ijhm.v1i2.127
18. Keshta, I., Odeh, A.: Security and privacy of electronic health records: Concerns and challenges. Egypt. Inform. J. **22**(2), 177–183 (2021). https://doi.org/10.1016/j.eij.2020.07.003

19. Paul, R.U., Mbeledogu, N.N., Sunday, B.T., Mbonu, C., Nwokoye, C.: A Web-based information management system for educational institutions in riverine areas. **4**(4), 31–40 (2022)
20. Van Damme, S., Sameri, J., Schwarzmann, S., Wei, Q., Trivisonno, R., De Turck, F., Torres Vega, M.: Impact of latency on QoE, performance, and collaboration in interactive Multi-User virtual reality. Appl. Sci. **14**(6), 2290 (2024). https://doi.org/10.3390/app14062290

Development and Testing of a Home-Based Phototherapy System for Neonatal Jaundice

Farah Amirah Azman, Rania Hussein Al-ashwal, Tian Swee Tan, and Maheza Irna Mohamad Salim

Abstract Neonatal jaundice is affecting up to 60% of newborns in their first week of life and its prevalence in Malaysia is about 301,000 cases annually. Jaundice is mainly caused by hyperbilirubinemia and can potentially lead to severe complications if untreated. Traditional hospital-based phototherapy, the standard treatment offered for jaundice, often lacks precision in irradiance exposure and disrupts mother-infant bonding due to the need for hospitalization. Therefore, this study proposes an innovative home-based phototherapy system with automatic light adjustment mechanisms to address these issues. The methodology started with the conceive phase which involves identifying user needs through benchmarking of the existing devices, brainstorming and conducting literature review related to the overall concept of the phototherapy system such as selection of components, irradiance levels, distance requirement of the light source and Total Serum Bilirubin monitoring. It is followed by the design and implementation. Finally, the testing phase was conducted to validate the system's performance through irradiance testing, evaluation of system responsiveness and conducting task analysis with parents and caregivers. This study has successfully developed a home phototherapy system with safety mechanisms like exposure duration limits and an ultrasonic sensor to maintain correct light distance. It uses blue LEDs for accurate irradiance and is monitored by a lux meter. This system not only meets the clinical criteria for phototherapy but also enhances treatment precision, supports mother-infant bonding by reducing hospital stays, and lowers healthcare costs, offering a safe and convenient alternative for managing neonatal jaundice at home.

F. A. Azman · R. H. Al-ashwal · M. I. M. Salim (✉)
Diagnostic Research Group, Health and Wellness Research Alliance, Building V01, Universiti Teknologi Malaysia, Johor Bahru, Johor, Malaysia
e-mail: maheza@biomedical.utm.my

T. S. Tan
Bioinspired Device and Tissue Engineering Research Group, Department of Biomedical Engineering and Health Sciences, Faculty of Electrical Engineering, Universiti Teknologi Malaysia, Johor Bahru, 81310 UTM Johor, Malaysia

IJN-UTM Cardiovascular Engineering Centre, Institute of Human Centered Engineering, Universiti Teknologi Malaysia, UTM Johor Bahru, 81310 Johor, Malaysia

K. W. Lai and P. S. Q. Yeoh (eds.), *Biomedical Engineering*, Series in BioEngineering,
https://doi.org/10.1007/978-981-97-9294-8_2

Keywords Irradiance · Jaundice · Phototherapy · Total serum bilirubin

1 Introduction

Hyperbilirubinemia is a common disease which arises when the liver fails to process bilirubin, a yellow-orange pigment found in the eyes and skin. Bilirubin can be found inside a bile, a fluid which the liver makes to digest food. Hence, phototherapy is a non-invasive treatment for neonatal jaundice which is very easy to apply and lowers total serum bilirubin (TSB) by inducing the creation of water-soluble bilirubin photo isomers that can be eliminated in urine and bile without the need for conjugation [1]. This treatment involves exposing the baby's skin to specific wavelengths of light, which helps convert bilirubin into a form that the liver can readily eliminate and the body is subsequently free from it. It operates by using light energy to change the insoluble and indirect bilirubin into soluble and direct bilirubin that may be excreted through the urine and feces by employing blue light sources such as light-emitting diodes (LEDs), fluorescent light bulbs, halogen lamps, sunlight, or others [2].

Current limitations of existing phototherapy devices include the need for constant monitoring in hospital-based settings to ensure proper irradiance and skin exposure [3]. Other than that, fiber optic based phototherapy devices are less effective, suffer from lower light intensity and often require additional phototherapy [4]. Conventional phototherapy which uses fluorescent lamps also suffers from reduced light emission over time and necessitates frequent maintenance [5]. Lastly, double-surface intensive phototherapy (DSIP) offers better outcomes by treating both surfaces of the baby simultaneously, but these advancements still require further development for effective home-based use [6].

The effectiveness of phototherapy depends on several aspects which are, the wavelength, distance between light source and neonates, irradiance level, as well as the body surface exposure. According to a conducted study, the use of light with a wavelength between 460 and 490 nm and a spectral irradiance of at least 30 μW/cm2/nm can maximize treatment in lowering blood bilirubin levels [7].

Home-based phototherapy has emerged as an alternative treatment option particularly for low to mid-level jaundice cases, due to its potential to maintain maternal-infant bonding, reduce hospitalization costs and provide continuous breastfeeding [5, 8]. Home phototherapy systems have been practiced in countries like Iran, Malaysia, Australia, the United States, and Britain [9]. These systems must adhere to guidelines specifying minimum irradiance levels to ensure efficacy.

This study aims to develop a home-based phototherapy system with an automated irradiance light mechanism tailored to clinical guidelines. The system will include electronic components such as ultrasonic sensors to automatically set the treatment distance from the total serum bilirubin input, LEDs, and a user interface for monitoring and adjustments for ensuring an accurate and effective treatment. Other than that, the double surface phototherapy design will also be employed for two light sources from top and bottom surface. The project will follow the concept of Conceive-Design-Implement-Operate (CDIO) framework in creating a portable, user-friendly, and efficient phototherapy device. The LED light sources will be chosen for their effectiveness in treating jaundice and their benefits over traditional light sources, such as lower energy consumption and longer lifespan.

The system's performance will be tested using task analysis and an irradiance meter to ensure accurate light exposure. Task analysis will involve collecting detailed user feedback to refine the system's usability and functionality. Optimization will address system performance, ensuring reliable operation and accuracy. The project's significance lies in providing a practical solution for parents to treat moderate neonatal jaundice at home, promoting mother-infant bonding and reducing healthcare costs.

2 Materials and Methods

2.1 Project Workflow

The project's workflow has been divided into 2 main stages. The first stage focuses on the development of a home phototherapy system with an automatic light control mechanism that meets clinical standards as shown in Fig. 1. It begins with the conceive phase which involves identifying user needs through benchmarking of the existing devices, brainstorming and conducting literature review related to the overall concept of the phototherapy system and followed by setting clear objectives that are aligned with the project's vision. The brainstorming sessions and a comprehensive literature review were conducted to generate innovative ideas and identify the current needs from clinical aspects, materials, and components. This stage is concluded with designing the electronic and mechanical aspect of the system and implementing the prototype development.

The second stage is testing and validating the developed system as shown in Fig. 2 below. It was started with strength simulations to ensure the structure can withstand infant loads up to 6 kg. The simulation result is used to further improve the mechanical design of the finalized prototype. Other than that, irradiance exposure is also monitored using a lux meter to ensure accurate light intensity, and system performance is optimized by evaluating responsiveness and identifying errors. Finally, user experience surveys are conducted with parents and caregivers to assess the system's usability, leading to the production of a finalized prototype.

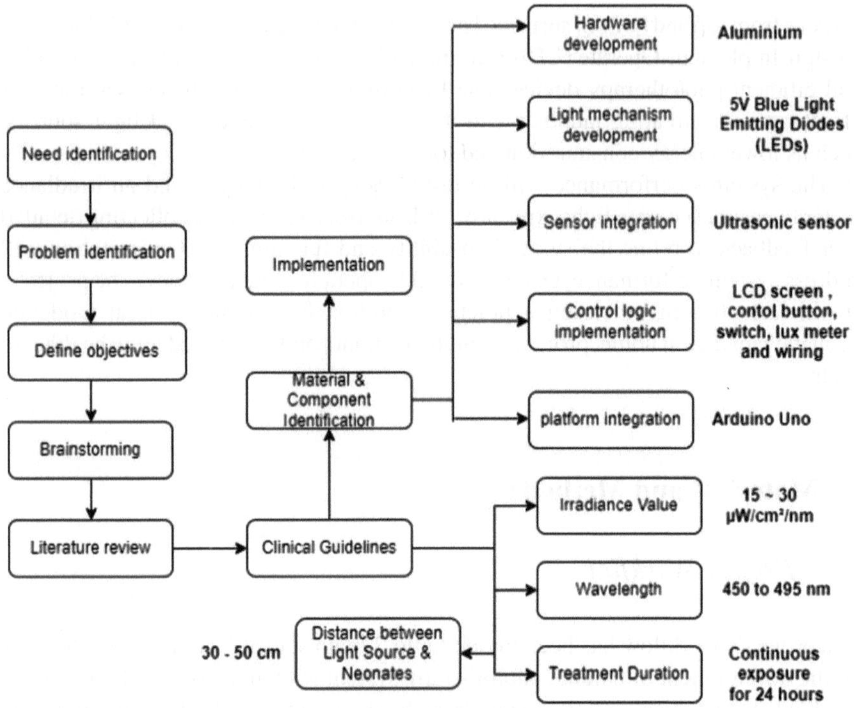

Fig. 1 Flowchart of the process in achieving the first objective of the project

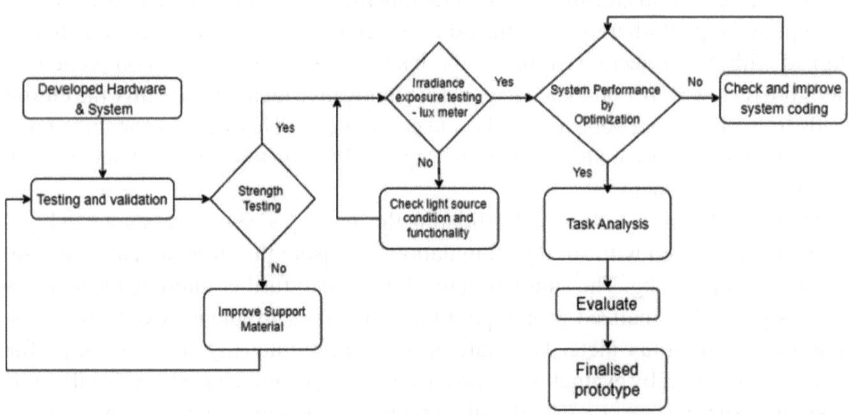

Fig. 2 Flowchart of the process in achieving the second objective of the projec

2.2 Phototherapy System Architecture and Operational Parameters

Figure 3 shows the top level architecture of the home based phototherapy system. It is designed specifically for the treatment of low and intermediate jaundice level. For the safety of treatment at home, the user must first input the Total Serum Bilirubin (TSB) value of 6 to 18 mg/dL, indicating low to intermediate jaundice suitable for home phototherapy [10]. If the TSB value falls outside this range, the system will alert the user and prevent the operation to ensure patient safety for home treatment as high TSB reading is dangerous and requires intensive treatment at the hospital. The system will automatically set the phototherapy duration based on TSB levels of 12 h/day for values below 10 mg/dL and 24 h/day for values above 10 mg/dL [10, 11]. The system also includes a pause and resume feature to allow for interruptions like diaper changes and breastfeeding to ensure accurate treatment duration. An ultrasonic sensor will monitor the distance between the light source and the neonate to maintain an effective range of 30 to 50 cm as stated by the Ministry of Health. The blue LEDs provide an irradiance of 15 μW/cm2/nm and are continuously monitored by a lux meter for accuracy.

2.3 Structural Components

Selecting appropriate materials for structural development is crucial to ensure that the phototherapy system is strong, durable, and capable of supporting patient weight during treatment. The base must be robust to support various loads, including the load from the components, infant bassinet, and the patient themselves. Additionally, the base should be broad and stable to prevent tipping during movement. Aluminum is chosen for the structure, particularly for the base and stand, due to its strength, corrosion resistance, and low reactivity to sterilization. Its lightweight yet strong nature makes it ideal for a home-based phototherapy device, ensuring it is easy to move, detach, and store. Aluminum material also allows for the addition of adjustable features, such as lock holes and slots easily without altering its properties. These features enhance safety by preventing accidental drops of the upper part of the structure.

The infant bassinet as shown in Fig. 4 is made from high-quality transparent plastic. This would allow the blue light exposure to reach the patient effectively while enabling parents or caregivers to observe the patient's condition easily. The 1.5 mm thick plastic also ensures durability and protection against accidental impacts and provides a secure environment for the neonates without adding excessive weight to the base structure. Hence, the infant bassinet design will be the same as the current design that most of the healthcare has been used.

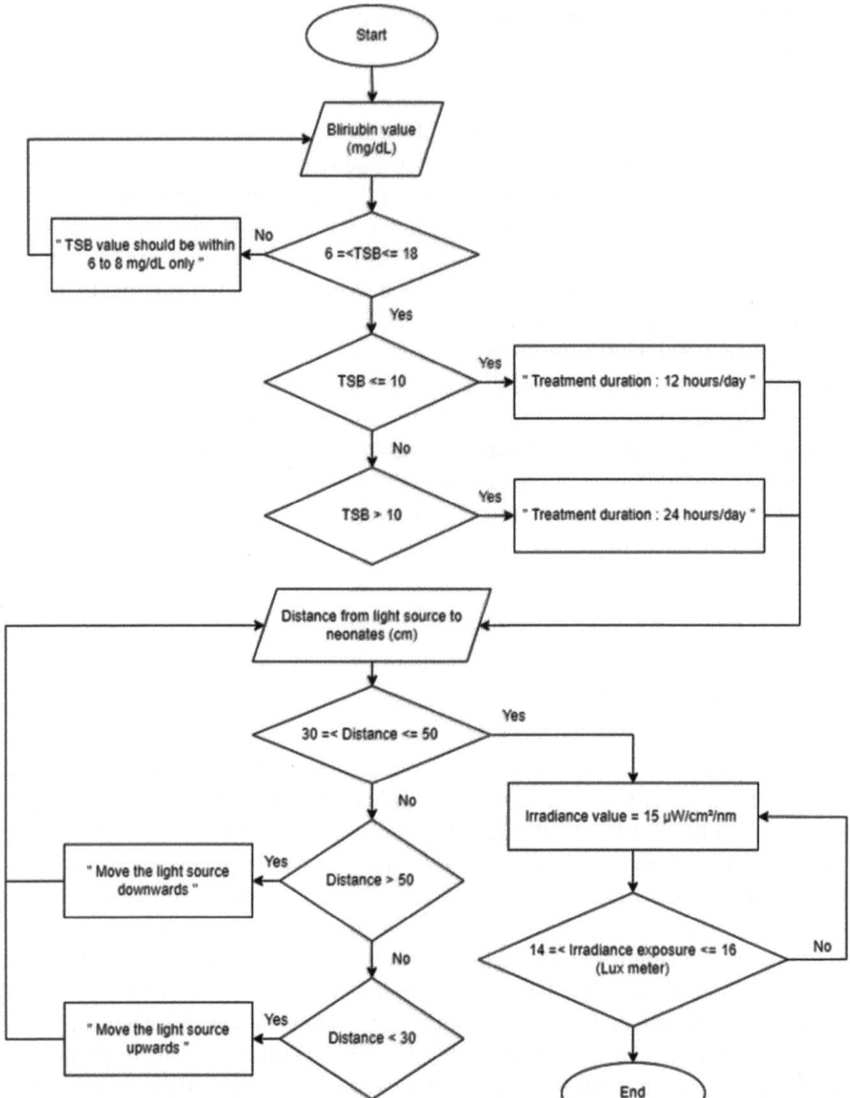

Fig. 3 Top level architecture of the system for home-based phototherapy

2.4 Electronic Key Components and Circuit Design

Several key components for the circuit connections of the phototherapy system were identified. A 5 V blue LED strip was chosen as the light source due to its ability to emit therapeutic wavelengths (450–495 nm) and its low power consumption. The Arduino Uno will serve as the central processing unit to control the system's functionality and

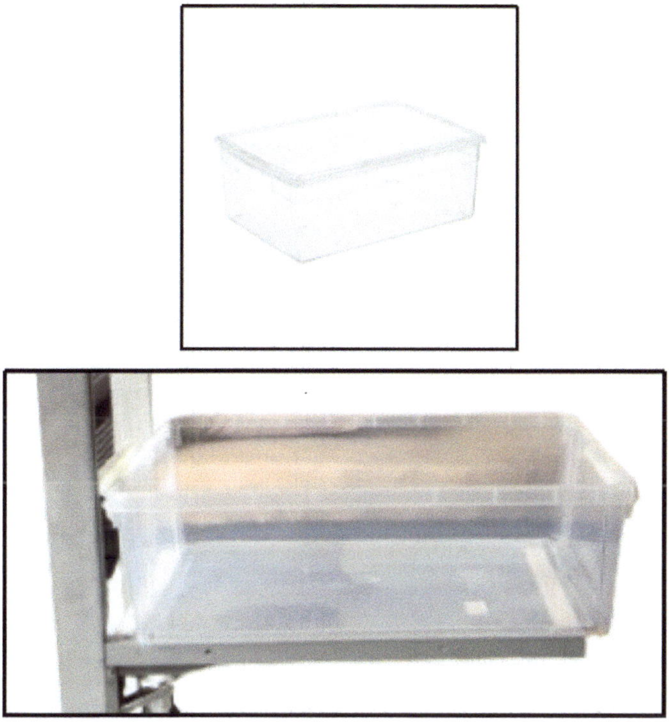

Fig. 4 Transparent plastic bassinet

the integration between the light source, timer, and ultrasonic sensor. The ultrasonic sensor is used to monitor the distance between the light source and neonates, ensuring an optimal height of 30–50 cm. For the user interface, an LCD screen and four buttons is used to set and display the Total Serum Bilirubin (TSB) values, height, and treatment duration for providing a user-friendly experience.

3 Result and Discussions

3.1 Hardware Assembly and Integration

The hardware development began with designing the model in SolidWorks, The focus of this stage was to achieve precise dimensions to ensure functionality and stability as in Fig. 5. The primary material used for the prototype is aluminum for its strength and lightweight properties. The design included various components such as screws for assembly and L brackets for detachable features. This allows the whole structure to be folded to improve portability. The first step in the assembly process was preparing

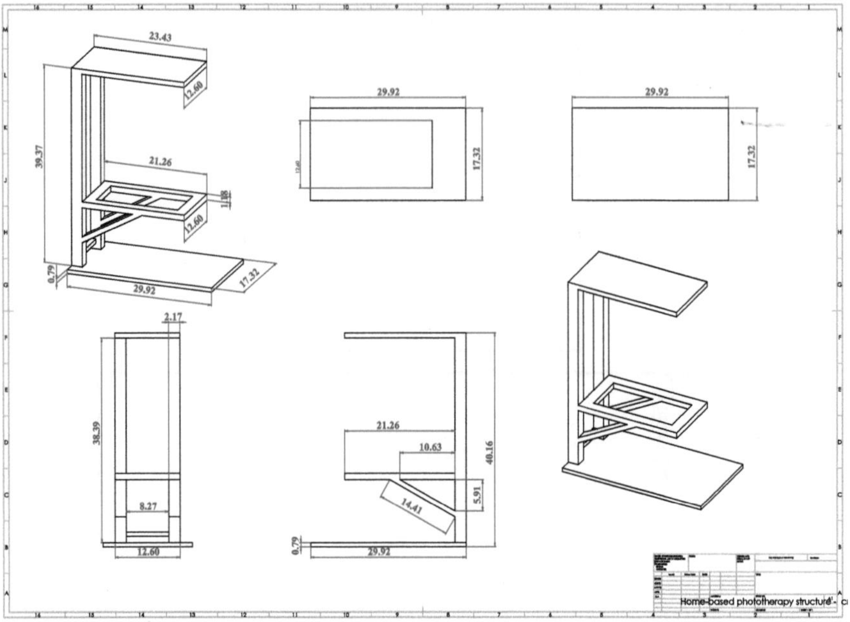

Fig. 5 Solidwork model and dimensions

the materials which involved cutting the aluminum to the specified dimensions. Each piece of the aluminum was carefully measured and marked before cutting to avoid errors and deburred to remove any sharp edges. This process is important to improve safety and ensure a better fit between components by eliminating burrs that could interfere with assembly.

Following the primary assembly, the hardware development proceeded with further detailed processes such as drilling to create slots in the aluminum where screws and other fastening components could be inserted. The L brackets were used on the detachable parts. These brackets were modified to allow for easier assembly through the pre-drilled slots. The modifications included adjusting the size and shape of the brackets to ensure a snug fit and ease of use. The addition of detachable features was a critical aspect of this research. These features are designed to make the system suitable for home use by allowing for easy storage and portability. The design as shown in Fig. 6 allows the system to be disassembled and reassembled without difficulty, less bulky and can be transported from one place to another.

Figure 7 shows the prototype structure after assembly. The structure was then adjusted to maintain alignment and stability of the prototype resulting in a robust system ready for testing and further optimization.

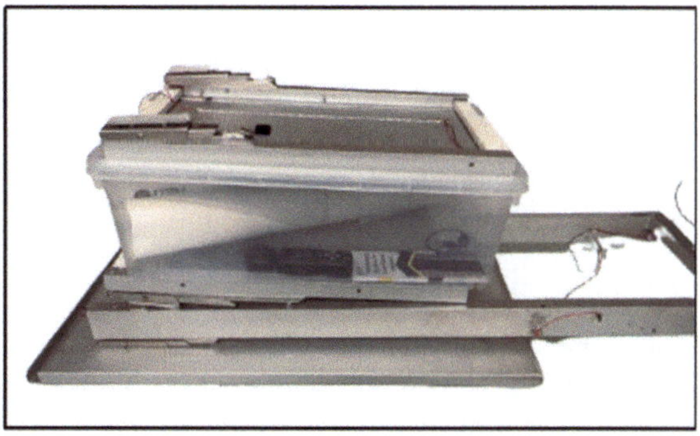

Fig. 6 Travel-friendly design and ease of storage

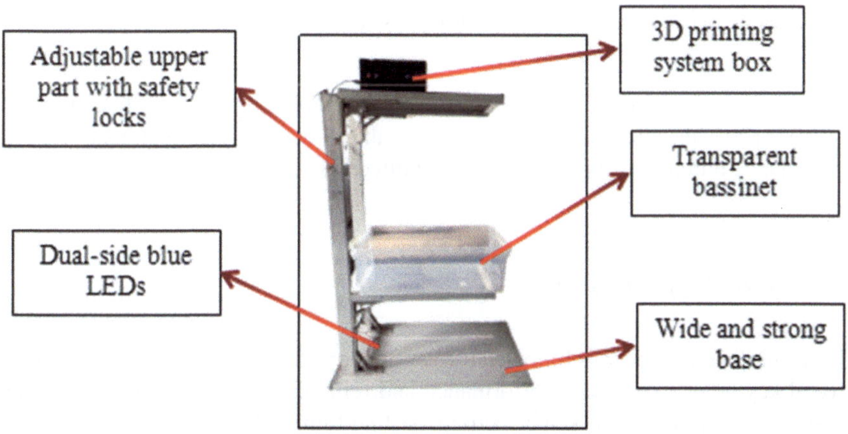

Fig. 7 Home based phototherapy system

3.2 Electronics System Integration

The electronics system integration stage began with designing the circuit and programming in Tinkercad to identify potential issues in the early stage of the development process before moving on to physical implementation. Once the simulation had confirmed the system worked as intended, the programming code was converted to Arduino IDE software. The code was then run and tested to optimize the system performance. After that, the code was transferred from Arduino IDE to the microcontroller. Simultaneously, circuit development was carried out which involved assembling hardware components, including sensors, LEDs, and other electronic parts, onto a breadboard.

The final step involved uploading the program to the Arduino Uno and making minor modifications based on the available components. These adjustments were made to fine-tune the system's functionality and ensure all parts worked harmoniously as planned in Fig. 8.

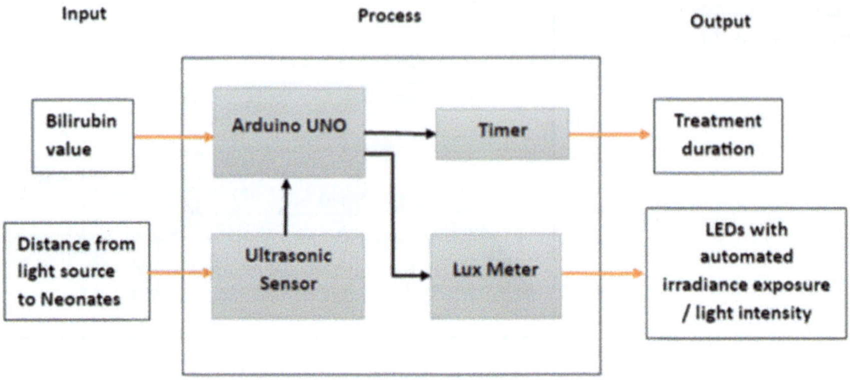

Fig. 8 Block diagram of home-based phototherapy system

After the integration of the electronic system and blue light source from top and bottom surface, the phototherapy system is capable of determining treatment duration and distance automatically based on the inputted TSB value.

3.3 System Optimization

The development stage was followed by a system optimization stage which involved several key steps to enhance performance and reliability of the system. The initial phase of the optimization stage focused on fine-tuning the programming code to reduce latency and improve response times by optimizing the algorithms. This is crucial to ensure effective phototherapy treatment. The user interface was also tested with multiple inputs to confirm its responsiveness and error-free operation. The next optimization step was light intensity adjustments through calibration of light source to maintain stable irradiance levels. The power supply and cooling mechanisms of the system were also optimized to prevent overheating as shown in Fig. 9. The casing which uses a 3D-printed box was also designed to be capable of supporting heat dissipation.

Other than that, ultrasonic sensor calibration which determines the distance between light source and neonates during treatment was also conducted to observe its accuracy. The measurements from the ultrasonic sensor were compared with measuring tape as shown in Figs. 10, 11 shows the calibration result of the ultrasonic sensor.

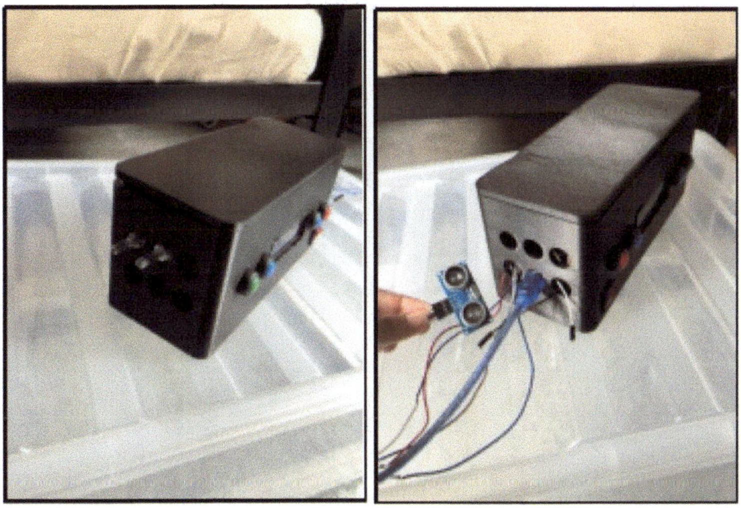

Fig. 9 Cooling mechanism

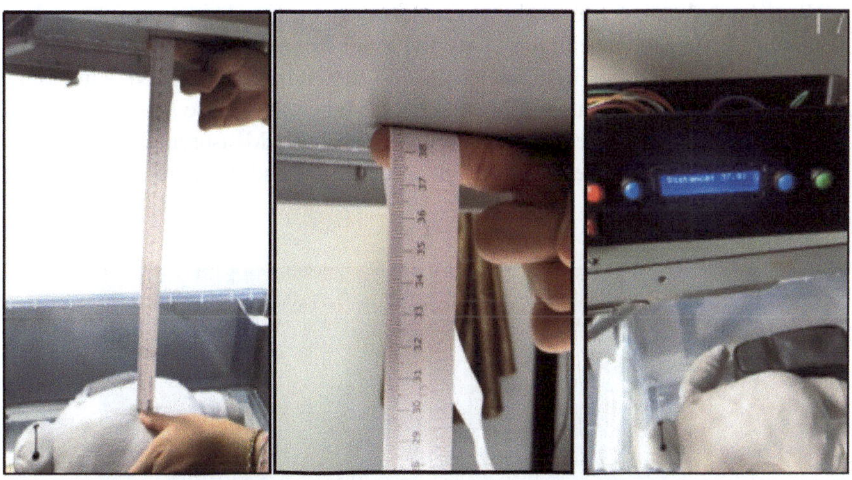

Fig. 10 Ultrasonic sensor calibration for treatment distance

3.4 Performance Testing and Analysis

3.4.1 Irradiance Exposure Using Lux Meter

The irradiance levels of the home-based phototherapy system were measured using a lux meter to ensure that the device meets the required irradiance levels specified by clinical guidelines. The lux meter measures the intensity of light emitted by the

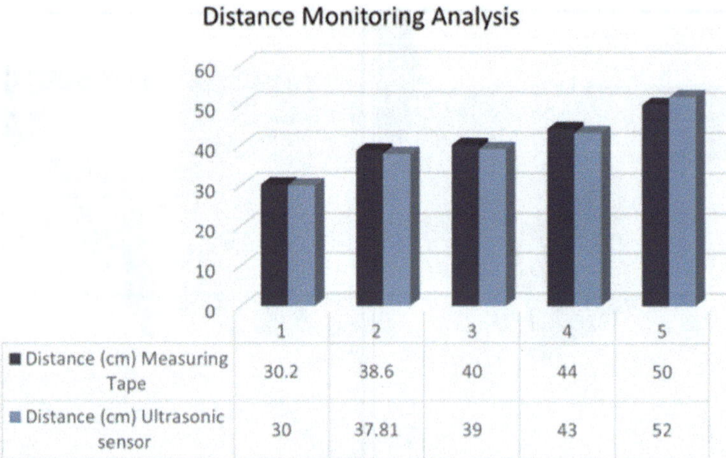

Fig. 11 Treatment distance comparison for ultrasound sensor calibration

phototherapy system in units of micro-watts per square centimeter per nanometer ($\mu W/cm^2/nm$). Irradiance level plays a crucial role in the jaundice treatment to reduce bilirubin value in neonates. In this study, the irradiance test was conducted in a dark room to eliminate external light interference as shown in Fig. 12. Then, a lux meter was placed at fixed points under the blue lights. When the device was switched on, the lux meter readings were recorded over a few seconds. Measurements were taken at various points to ensure uniform light distribution and accurate assessment of irradiance levels.

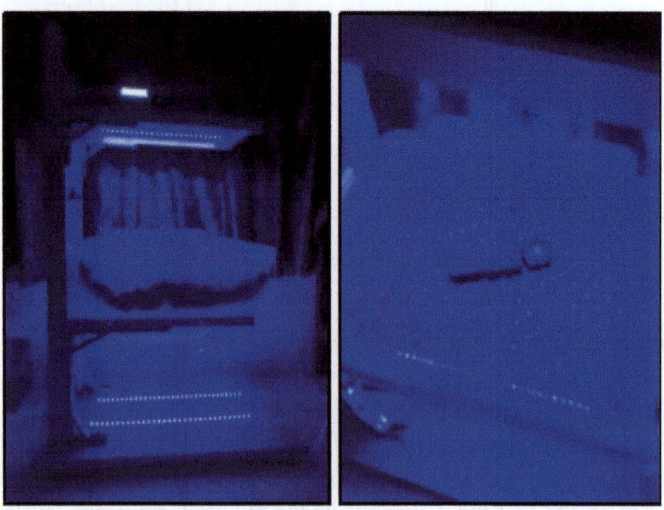

Fig. 12 Measuring light source irradiance in controlled environment

Throughout the test, it was observed that uniform light distribution was confirmed across the targeted treatment area. The system is able to achieve the required irradiance of 15 $\mu W/cm^2/nm$. The consistent and adequate irradiance levels indicate that the home-based phototherapy system provides the necessary therapeutic light exposure for effective treatment. The uniformity of light distribution ensures comprehensive treatment coverage, which is essential for reducing bilirubin levels in neonates.

3.4.2 Load Simulation Testing

Load simulation tests were conducted by using solidwork to evaluate the maximum capacity of the phototherapy system under different load conditions. This testing is crucial to ensure the system's robustness and durability by subjecting the phototherapy system to various load conditions ranging from 3 kg up to 8 kg in accordance with the average weight of neonates [12]. Load or force in downward direction indicated by purple arrow were applied on the bassinet as shown in Fig. 13 for extended period of usage and the resulting displacements were recorded.

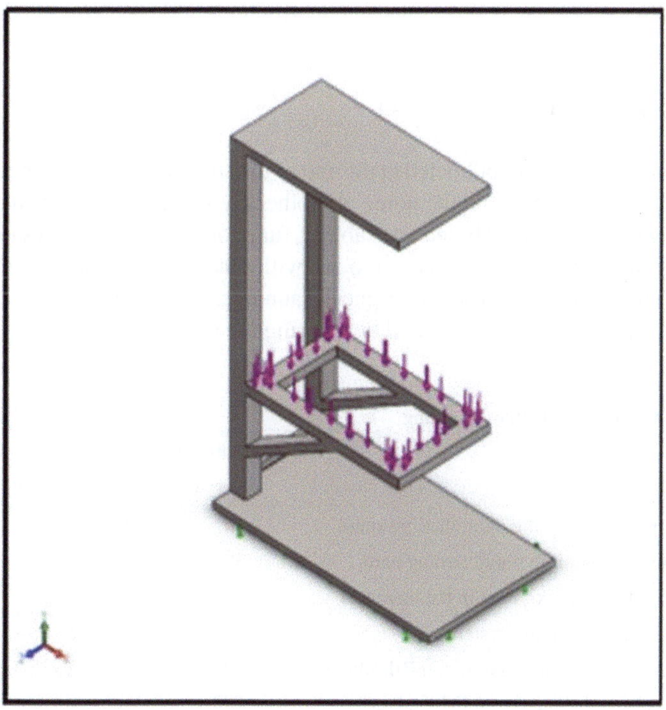

Fig. 13 Downwards force/load were applied on bassinet area

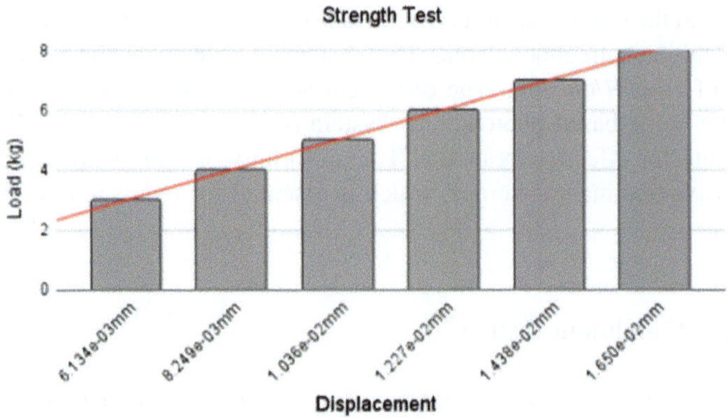

Fig. 14 Bar chart of displacement vs load (kg)

The result for load simulation test shown by Fig. 14 demonstrates that the system remained stable when subjected to loads up to 4 kg throughout an extended continuous usage period. However, fracture was observed to occur at the bottom part when loads of 6 kg and above were applied to the structure. The structure also exhibited slight bending with increasing displacement with higher loads.

3.4.3 Task Analysis

Task analysis survey was conducted at the end of the study to provide comprehensive insights into user feedback on the home phototherapy system [13]. The survey tasks conducted were categorized based on usability, functionality, and overall satisfaction. The survey included 10 users aged 20 to 50, with 60% being mothers or caretakers and 60% having children who experienced jaundice. Table 1 shows the results of respondent's rating for each assigned task during the evaluation.

Table 1 Summary of task analysis rating

Task	Rating (%)			
	Very easy	Easy	Neutral	Difficult
System setup: Setting up the phototherapy stand	40	30	30	0
System setup: Assembling phototherapy parts	30	20	40	10
System setup: Placing 4 kg load in the bassinets	30	50	20	0
Operation: Switching on the system	40	40	0	10
Operation: and setting the Total Serum Bilirubin reading	40	40	20	0
Operation: Setting distance from light source	100	0	0	0
Operation: Treatment duration and light activation	100	0	0	0

The first task given to the respondents during the task analysis session was to set up the phototherapy system stand from its folding stage. All respondents successfully assembled the stand with 40% of the respondents rated the task as "very easy," 30% "easy," and 30% "neutral". Additionally, some of the respondents commented that the slots for assembling the parts were not always smooth and hence required additional force for the setup. The next task was assembling the phototherapy parts which involved connecting wires from the light source to the processor box. 90% of the respondents successfully assembled all parts with 50% of them rated the task as easy and very easy. For placing a 4 kg load into the bassinet, the majority of the respondents agreed that the task was easy. Additionally, half of the respondents felt it was safe and easy to place neonates inside the bassinet whilst 70% found the structure was sturdy enough to hold their neonates. Under the system operation tasks, 90% of respondents successfully switched on the system, with one failure potentially due to unstable wire connections. All respondents were also able to input the TSB values correctly indicating the user-friendly interface. For height adjustments, respondents found the system's instructions were helpful, though some noted issues with slot smoothness. All respondents confirmed the system's functionality, including automatic light activation and countdown operations, with half finding the system starts treatment quickly after adjustments. The overall experience was rated "Excellent" by 60% of respondents, "Good" by 20%, and "Neutral" by the remaining 20%. This feedback highlights the strengths in user interface design and suggests areas for improvement in the physical assembly process. These insights are crucial for guiding future enhancements to increase user satisfaction and system reliability.

4 Conclusions

The development of a home-based phototherapy system with automated light adjustment and treatment time is a significant advancement in neonatal jaundice treatment. Rigorous testing confirmed the system's compliance with clinical standards and user-friendliness, making it a promising alternative to hospital-based phototherapy. This innovative approach can reduce hospital stays and lower the risk of infections. It also decreases healthcare costs while enhancing mother-infant bonding. The system ensures consistent and optimal light exposure, reducing human error and improving treatment effectiveness. However, limitations include material strength for infants over 4 kg and assembly difficulties due to unsmooth slots and wire connections. Future improvements should focus on using stronger, corrosion-resistant, lightweight materials, simplifying assembly, and incorporating jaundice meters for real-time monitoring. Conducting large-scale usability testing with telemedicine support is also recommended. Addressing these issues will further enhance the system's effectiveness, usability, and user satisfaction, transforming neonatal care by providing a safe, effective, and economical home treatment option.

References

1. Mreihil, K., Šaltytė Benth, J., Stensvold, H.J., Nakstad, B., Hansen, T.W.R.: Norwegian NICU phototherapy study group, & norwegian neonatal network. Phototherapy is commonly used for neonatal jaundice but greater control is needed to avoid toxicity in the most vulnerable infants. (2020)
2. Ministry of Health Malaysia.: Management of neonatal hyperbilirubinemia: health technology assessment report. Health Technol. Assess. Unit, Med. Dev. Div., Minist. Health Malays. (2002)
3. Zanardo, V., Guerrini, P., Sandri, A., Ramon, C.M., Severino, L., Garani, G., Mesirca, P., Straface, G.: Pilot study of home phototherapy for neonatal jaundice monitored in maternity ward during the enforced Italy-wide COVID-19 national lockdown. Eur. J. Pediatr. **181**(9), 3523–3529 (2022). https://doi.org/10.1007/s00431-022-04557-x
4. George, P., Lynch, M.: Ohmeda Biliblanket vs Wallaby phototherapy system for the reduction of bilirubin levels in the home-care setting. J. Neonatal Nurs. **24**(5), 330–336 (2018)
5. Chua, B.S., Song, L.H., Chang, C.T., Lim, X.J., Nachiappan, J.: Drive-through transcutaneous bilirubin screening for neonatal jaundice: A safe and efficient system during the COVID-19 pandemic. Zenodo (2021). https://doi.org/10.5281/zenodo.7244152
6. Sabzehei, M.K., Waisi, S., Shokouhi, M., Tapak, L.: Single-surface intensive phototherapy or double-surface intensive phototherapy in neonatal non-hemolytic hyperbilirubinemia: A comparison of effectiveness and complications. Med. J. Islam. Repub. Iran, **35**, 192 (2021). https://doi.org/10.47176/mjiri.35.192
7. Mitra, S., & Rennie, J. (2017). Neonatal jaundice: aetiology, diagnosis, and treatment. Br. J. Hosp. Med. (Lond.), **78**(12), 699–704. https://doi.org/10.12968/hmed.2017.78.12.699
8. Namnabat, M., Mohammadizadeh, M., Sardari, S.: The effect of home-based phototherapy on parental stress in mothers of infants with neonatal jaundice. J. Neonatal Nurs. **24**(5), 323–329 (2018). https://doi.org/10.1016/j.jnn.2018.09.001
9. Snook, J.: Is home phototherapy in the term neonate with physiological jaundice a feasible practice? A systematic literature review. J. Neonatal Nurs. **23**(1), 28–39 (2017). https://doi.org/10.1016/j.jnn.2016.08.001
10. Ministry of Health Malaysia.: *Clinical Practice Guidelines: Management of Neonatal Jaundice (Second Edition)*. Malays. Health Technol. Assess. Sect. (MaHTAS), Minist. Health Malays. (2015). Available at http://www.moh.gov.my
11. Sroufe, N.S., Vredeveld, J.L., Goodson, S.L., et al.: Management of indirect neonatal hyperbilirubinemia. Mich. Med. Univ. Mich., Ann Arbor, MI. (2020). Available at https://www.ncbi.nlm.nih.gov/books/NBK567493/
12. Boo, N.: Outcome of very low birthweight neonates in a developing country: Experience from a large Malaysian maternity hospital. Singap. Med. J., **33**(1), 14–18 (2019). Retrieved from http://smj.sma.org.sg/3301/3301a4.pdf
13. Team, Ux.: UX task analysis: A complete guide + example. (2023). Retrieved from https://blog.uxtweak.com/ux-task-analysis/

Non-invasive Blood Glucose Monitoring via Infrared Absorbance

Mohamed Ahmed Mohamed Mohamed Saleh, Asnida Abdul Wahab, Muhamad Amin Abd Wahab, Anas Mohd Noor, and Muhammad Faiz Md Shakhih

Abstract Diabetes affects millions globally, requiring consistent monitoring of blood glucose levels (BGL) to prevent health complications. Currently, glucometers and glucose patches are the most effective monitoring technologies, but they are invasive or minimally invasive, which can be cumbersome, especially for children and the elderly. This has led to a growing demand for non-invasive glucose monitoring methods, which have become the focus of much research. Despite the emergence of promising techniques, challenges related to practicality and effectiveness have hindered their adoption. This study aims to validate near-infrared absorption (NIR) as a non-invasive glucose monitoring method. A prototype device was developed using a 940 nm infrared (IR) emitter and a photodiode detector with a detection range from 800 to 1600 nm. The prototype was tested for sensitivity using solution samples with varying glucose concentrations and for selectivity using common blood components like ascorbic acid, tartaric acid, and lactic acid mixed with glucose solutions. Initial tests without optimizations showed a good correlation with R^2 value of 0.85. With optimizations, the correlation for the sensitivity improved to 0.87. These findings suggest that NIR has the potential to be a viable non-invasive glucose monitoring method, paving the way for further research and development of practical devices for everyday use.

Keywords Glucose sensor · Near infrared · Diabetes · Non-invasive monitoring

M. A. M. M. Saleh · A. A. Wahab · M. A. A. Wahab · M. F. M. Shakhih (✉)
Department of Biomedical Engineering and Health Sciences, Faculty of Electrical Engineering, Universiti Teknologi Malaysia, Johor, Malaysia
e-mail: mfaizms@utm.my

A. A. Wahab · M. F. M. Shakhih
Advanced Diagnostics and Progressive Human Care Research Group (DIAGNOSTICS), Faculty of Electrical Engineering, Universiti Teknologi Malaysia, Johor, Malaysia

A. A. Wahab
Medical Device and Technology Centre (MEDiTEC), Universiti Teknologi Malaysia, Johor, Malaysia

A. M. Noor
Faculty of Electronic Engineering & Technology, Universiti Malaysia Perlis, 02600 Arau, Malaysia

K. W. Lai and P. S. Q. Yeoh (eds.), *Biomedical Engineering*, Series in BioEngineering,
https://doi.org/10.1007/978-981-97-9294-8_3

1 Introduction

Glucose monitoring plays a vital role in the management of diabetes, requiring consistent and accurate tracking of blood glucose levels to maintain proper control and prevent complications [1]. Traditional glucose monitoring methods include the use of blood glucometers, which require a small blood sample typically obtained through finger pricking. Continuous Glucose Monitoring (CGM) systems represent a more advanced approach, utilizing sensors inserted beneath the skin to continuously measure glucose levels in the interstitial fluid. Additionally, innovations like Flash Glucose Monitoring have emerged, enabling users to scan sensors worn on the skin to access glucose data without the need for frequent blood samples [2]. While these technologies have significantly improved diabetes management, they remain invasive and uncomfortable, posing challenges for regular use, especially among non-diabetic individuals who could benefit from preventative glucose monitoring.

To address the limitations inherent in these traditional methods, researchers have been exploring non-invasive alternatives, with Infrared (IR) technology emerging as a promising option. One of the most notable approaches involves Near-Infrared Absorption (NIR Absorption), where light in the near-infrared spectrum is directed onto the skin. This method exploits the unique absorption patterns of glucose molecules, providing a potential means to estimate their concentration [3]. Spectroscopic analysis and calibration models, often enhanced by machine learning algorithms, are employed to correlate these patterns with glucose levels traditionally measured through invasive methods. However, the challenge lies in accurately isolating glucose-specific signals from the complexities of skin composition, where variables such as hydration, skin thickness, and the presence of other compounds can significantly affect the accuracy of the readings [4]. Despite these challenges, ongoing research continues to refine NIR-based glucose monitoring devices, aiming to improve their accuracy, reliability, and usability with the goal of developing a commercially viable and user-friendly product.

This chapter aims to establish the foundation for understanding the potential of NIR technology in non-invasive glucose monitoring. The study outlined here investigates the feasibility of utilizing a simple prototype model, incorporating widely available infrared emitters such as 940 nm LEDs, and laboratory samples of distilled water mixed with dextrose, which closely mimics glucose in blood. The focus of this investigation is on evaluating the sensitivity and selectivity of the technology, alongside exploring basic optimization methods to determine how slight changes in calibration can impact the accuracy of the measurements.

Accurate monitoring of glucose levels is essential not only for individuals with diabetes but also for patients who require regular glucose assessments in clinical settings. The importance of such monitoring extends beyond diabetes management to ensuring patient safety during routine checks and emergencies. Current methods, although precise, are often invasive and uncomfortable, necessitating frequent finger pricking or sensor insertions that can be burdensome for patients who require constant monitoring [5]. As technological advancements open new avenues for non-invasive monitoring, NIR absorption-based methods present a promising alternative. However, these methods are still in the developmental stage and face significant challenges, including interference from other substances and difficulties in calibration, leading to the potential for unreliable results [6]. Given these challenges, it is important to validate the accuracy and reliability of NIR-based glucose monitoring devices before they can be widely adopted. This study seeks to contribute to this validation process by investigating the dependability and precision of these devices under varying conditions, including the presence of different impurities and skin characteristics.

The primary objective of this study is to validate infrared absorption as a viable modality for glucose sensing, with a particular focus on sensitivity and selectivity. To achieve this, the research is structured around several key objectives: developing a prototype glucose monitoring device based on NIR technology, evaluating the correlation between voltage readings and glucose concentrations using the prototype and analysing the impact of various optimization techniques, such as using covers and dyeing samples with food colouring, on the performance of the prototype. This introduction provides an overview of the potential of NIR technology to revolutionize glucose monitoring, offering possibility into a future where non-invasive methods could become a standard approach to maintaining optimal health.

2 Literature Review

2.1 Blood Composition and Blood Glucose

It's important to study the anatomy of blood as it gives an idea of the medium that will be subjected to most tests in regard to glucose monitoring. While many other biological parameters outside of blood may affect the overall output of a device, non-are as volatile as blood in terms of changes to their structure and compositions over a relatively short period of time. Figure 1 shows a comprehensive overview on blood composition in the body.

The general composition of blood includes around 55% plasma and 45% blood cells that include red blood cells, white blood cells and platelets. Plasma alone is further separated into 92% water, 7% proteins and 1% nutrients and other solutes that include glucose. Effectively glucose concentrations comprise around 0.1% of total blood composition, while this value may seem small it is the bases for a large portion of the body's energy usage. Parameters like age, weight, gender and body composition (fat/muscle percentages) can affect glucose homeostasis. For example when looking at gender differences in glucose tolerance impairment, the prevalence of impaired fasting glucose (IFG) is higher in men, while impaired glucose tolerance (IGT) is more prevalent in women [7]. The reasons for these differences in early dysglycemia are unknown but could involve the effect of gonadal hormones. For

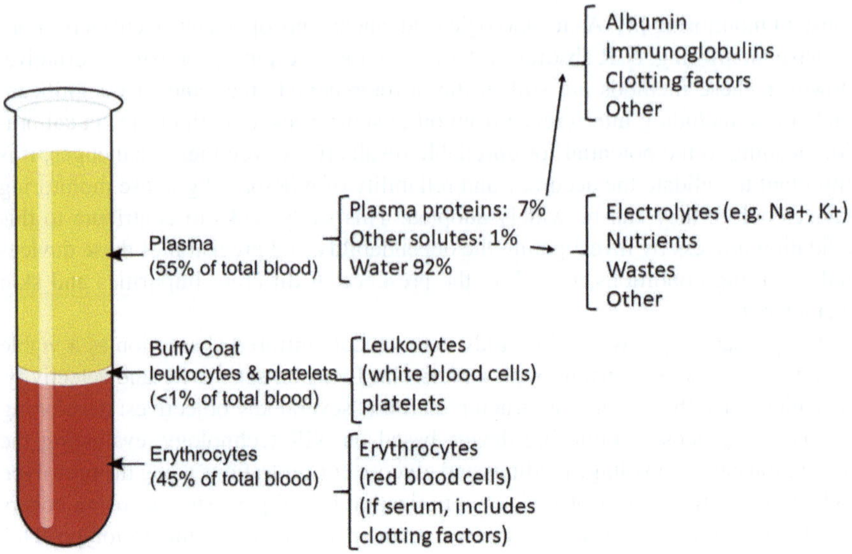

Fig. 1 Comprehensive overview of overall blood composition

example, menopausal hormone therapy with estrogens decreases fasting glucose while impairing glucose tolerance.

As shown in Table 1, glucose has a very large concentration in plasma relative to other components, collectively other protein components like Human Serum Albumin (HSA), Glutamine, Alanine, etc. collectively form a higher concentration in the plasma when compare with glucose. These components have a very different composition to that of glucose and thus can have a different reaction to specific wavelengths of infrared light.

In the context of near-infrared spectroscopy (NIRS) and imaging, chromophores in the blood primarily refer to molecules that absorb and interact with near-infrared light. Some common near-infrared chromophores in the blood include hemoglobin, separated into oxyhemoglobin (HbO2) and de-oxyhemoglobin (Hb). Oxygenated hemoglobin absorbs near-infrared light differently than deoxygenated hemoglobin,

Table 1 Concentrations of most common components in plasma [8]

Standard	Mean concentration (μM) in healthy adult males ± St. dev.
Glucose (fasting)	4545 ± 465
HSA	654 ± 67.9
Glutamine	645 ± 64
Alanine	419 ± 89
Valine	253 ± 37
Glycine	236 ± 43

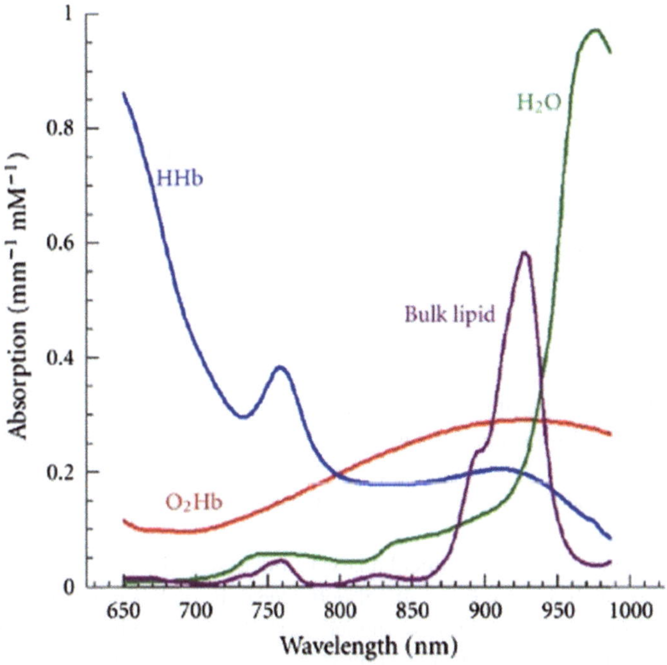

Fig. 2 Endogenous absorption related to wavelength [9]

absorbing less light at certain near-infrared wavelengths, meanwhile deoxygenated hemoglobin absorbs more near-infrared light, especially around 800 nm as illustrated in Fig. 2. We can also see that water is a relatively weak absorber of near-infrared light, but it can still contribute to the overall absorption spectrum. Cytochrome c oxidase an enzyme that is part of the electron transport chain in mitochondria and contains metal centers that can absorb near-infrared light. And lipids which are more commonly associated with absorption in the visible range, yet they can also contribute to absorption in the near-infrared range.

2.2 *Near Infrared Spectroscopy for Glucose Monitoring*

Near-Infrared Spectroscopy (NIRS) operates on the principle that molecules absorb light within the near-infrared region of the electromagnetic spectrum. This process is governed by key principles, including the concept of absorption bands, where infrared light interacts with molecular vibrations. These vibrations, including overtones and combinations of fundamental vibrations, lead to absorption bands that are characteristic of specific functional groups within molecules.

NIRS techniques rely on two main measurements which are transmittance and reflectance. Transmittance involves measuring the amount of light that passes through a sample, while reflectance measures the light reflected off a sample at different wavelengths. The intensity of the transmitted or reflected light provides valuable information about the sample's composition. A fundamental aspect of NIR absorption is its reliance on the Beer-Lambert Law, which describes the relationship between absorbance, concentration, and path length. This law is instrumental in correlating the intensity of light absorption with the concentration of specific components in a sample. The Beer-Lambert equation is expressed as:

$$A = \varepsilon c l$$

where A represents the absorbance, ε represents the molar absorption coefficient in (M^{-1} cm^{-1}), c Represents molar concentration in (M) and l represents optical length in (cm). This relationship can be visually represented, as shown in Fig. 3, which illustrates the effect of glucose concentration on the intensity of light passing through a medium.

NIRS is widely used for chemical analysis and concentration analysis of molecular structures within substances. Its effectiveness in quality control and analysis work is well-documented, with the added advantage of posing no inherent risk to the samples being tested. In the context of glucose monitoring, NIRS is being actively explored, particularly for non-invasive measurement of blood glucose levels. Researchers are investigating the potential of NIRS to provide real-time insights into glucose metabolism by tracking changes in tissue oxygenation and blood flow [11]. In clinical settings, NIRS is studied as a tool for continuous glucose monitoring, especially in scenarios where traditional methods, such as glucometer measurements, may be impractical or uncomfortable for patients [3].

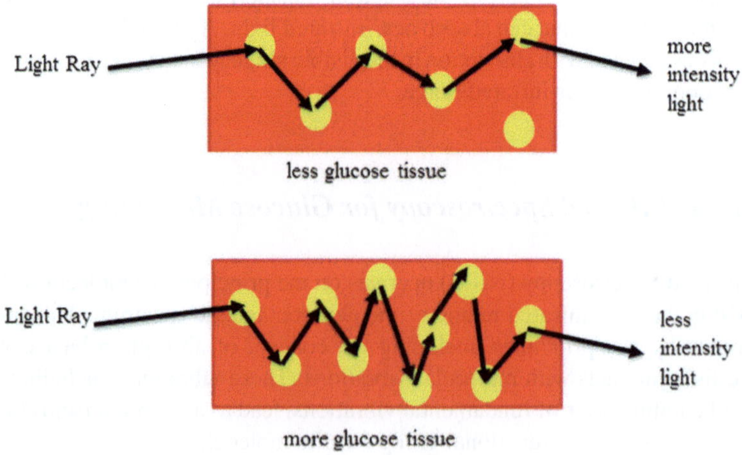

Fig. 3 The Effect of glucose concentration on light intensity [10]

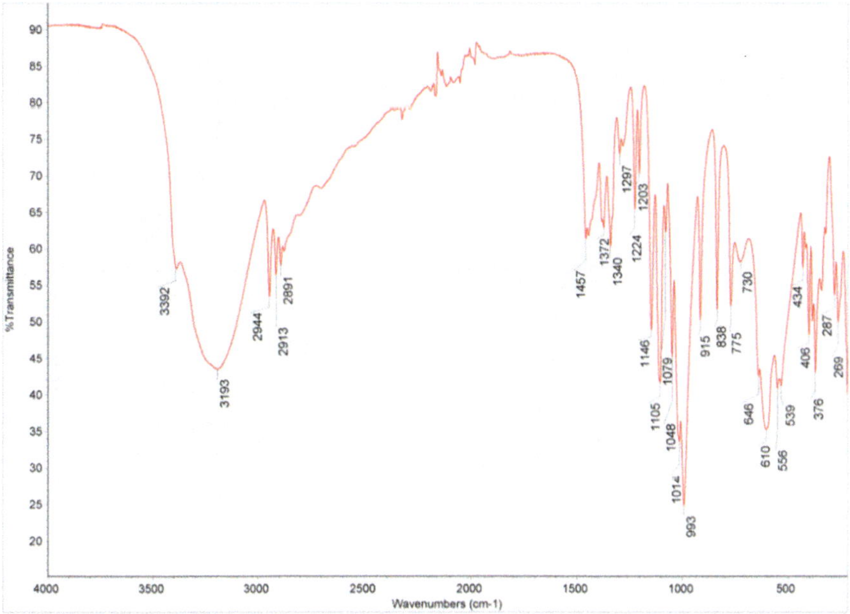

Fig. 4 Full infrared spectrum absorbance of glucose

Figure 4 illustrates the infrared transmittance of glucose, showing how NIRS can analyze near-infrared light absorption patterns to correlate these with variations in blood glucose levels. These patterns are related to known absorption characteristics of common components in blood.

While NIRS holds significant promise for glucose monitoring, ongoing research and development are essential to overcoming challenges related to calibration, accuracy, and the practical application of the technology across diverse physiological conditions. As advancements continue, integrating NIRS into wearable devices could further enhance its accessibility for continuous glucose monitoring and expand its applications in healthcare and biomedicine.

3 Research Methodology

3.1 Infrared Glucose Sensor Prototype Development

The prototype is built around an Arduino UNO R3 microcontroller, which serves as the central unit for powering components and processing data from the detector. A 940 nm IR emitter LED is paired with a photodiode receiver capable of detecting a wide range of wavelengths from 800 to 1600 nm. The choice of the 940 nm

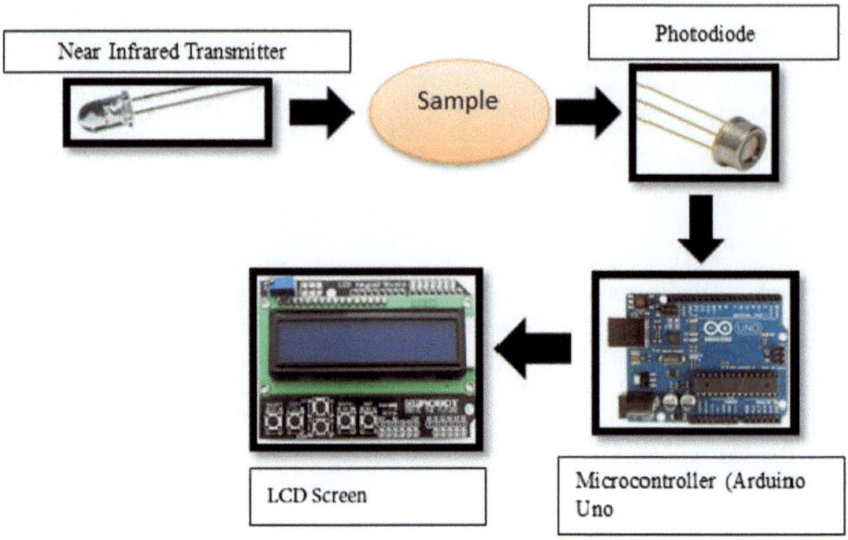

Fig. 5 Block diagram of the circuit components

wavelength for the emitter is based on its alignment with the absorption pattern of glucose, as compared to other biological parameters that might influence the signal. The selected wavelength ensures that the signal generated by glucose is strong enough for variations in sample concentration and composition to produce noticeable changes. This decision was informed by the absorption pattern of glucose illustrated in Fig. 4 of the literature review. Additionally, the 940 nm wavelength is the most commonly available IR emitter on the market, making it a practical choice for industrial production of a functional clinical product. While wavelengths such as 1090 nm and 1550 nm could potentially yield better results, they are considerably more expensive and difficult to source, which makes them less attractive for manufacturing purposes. If necessary, additional wavelengths may be incorporated to enhance the clarity of the results. The sample will be placed between the IR emitter and the photodiode receiver, with environmental controls implemented to maintain a consistent temperature around the samples, thereby minimizing any impact on the signal. Optionally, a 16×2 LCD screen could be integrated to display the voltage readings from the detector. The overall configuration of the system is depicted in the block diagram shown in Fig. 5.

3.2 Sensitivity and Selectivity Testing

The samples used for the testing process will be divided into two main categories: sensitivity testing and selectivity testing. Sensitivity testing focuses on the device's ability to detect the minimum input of physical parameters that will create detectable output change [1]. In contrast, selectivity testing assesses the device's effectiveness in maintaining a stable signal output when a sample with a constant glucose concentration is combined with other components that may introduce noise.

For sensitivity testing, the samples will consist of distilled water and dextrose sugar. A control sample of distilled water with no added dextrose sugar will serve as the baseline for voltage readings. Subsequent samples will have incremental additions of 50 mg of dextrose sugar per decilitre (dL), resulting in five samples with concentrations of 100, 150, 200, 250, and 300 mg/dL, or 5.6, 8.3, 11.1, 13.9, and 16.6 mmol/L, respectively. Dextrose sugar is chosen as a substitute for glucose due to its similar composition and interaction with infrared wavelengths. This selection of concentrations is designed to produce a comprehensive range of voltage readings based on varying glucose concentrations. Additional samples may be introduced to create a smoother correlation between concentration changes, depending on the results of initial tests. The primary goal is to correlate changes in glucose concentration with corresponding changes in voltage readings.

For selectivity testing, the focus will be on evaluating how the signal changes when different organic compounds are introduced to a single concentration of glucose diluted in distilled water. The chosen concentration is 100 mg/dL (5.6 mmol/L), which approximates the average blood glucose level in a healthy adult. Various samples will be prepared by adding different organic compounds such as ascorbic acid, tartaric acid, and lactic acid at concentrations similar to their median levels in a healthy male body. These compounds are commonly present in blood and can interfere with the signal due to their molecular structures being similar to glucose, increasing the likelihood of signal noise. The test will be repeated for each of the optimization methods implemented. The goal is to assess the noise-to-signal ratio and determine the level of noise that the signal can tolerate before becoming unreadable.

During testing, the samples will be placed between the IR emitter and the photodiode detector, as illustrated in Fig. 6. The process will begin with the control sample for each test parameter (0 mg/dL for sensitivity testing and 100 mg/dL for selectivity testing), using the control sample's signal reading as the baseline for comparison. Voltage readings from each sample will be taken to correlate changes in concentration and composition. The resulting data will be plotted to examine the strength of these correlations. Additionally, the effects of various optimization methods will be compared to evaluate their positive impact on the overall results and their statistical significance.

The photodiode will generate a voltage reading proportional to the infrared light absorbed by the sample at a specified exposure rate, typically around 10 ms. In this application, the photodiode's reading will directly correlate with the sample's absorption of near-infrared (NIR) light. Higher absorption of NIR light results in less IR light reaching the photodiode sensor, leading to a lower voltage reading. By analysing this relationship between IR light absorption and photodiode voltage output, it becomes possible to assess absorption differences across differently prepared samples.

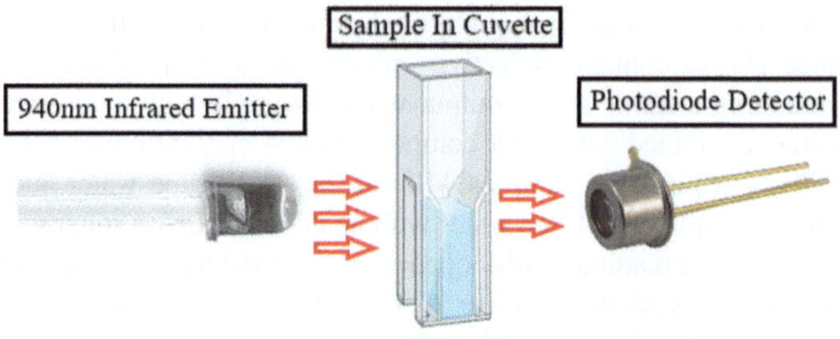

Fig. 6 Simplified model of the sample testing procedure using emitter and detector probes

3.3 *Optimization of Glucose Measurement Analysis*

A key aspect of the testing process involves optimizing both the environment and the samples to improve the accuracy and reliability of the readings. These optimization methods primarily impact two parameters: level change and overall sensitivity.

Level change refers to a consistent shift in all measurements by a similar amount (e.g., from 4-3 V to 3-2 V). This change can result from an increased overall absorption of infrared by the samples or reduced environmental noise affecting the detector. One way to achieve this is by enhancing the sample's ability to absorb more infrared rays. This can be done by adding commercial red food colouring that does not contain glucose, ensuring it doesn't alter the sample's composition. The red colouring modifies the sample's properties, making it more absorbent to near-infrared wavelengths while reducing its absorption of visible red wavelengths.

Overall sensitivity refers to the device's ability to detect changes in material concentration or composition. Essentially, it allows the device to register a greater voltage difference for the same change in concentration or composition. Sensitivity can be improved by reducing the scattering of infrared light after it interacts with the sample. This ensures that the infrared signals are primarily influenced by the sample's properties rather than external factors, such as the surrounding environment. To achieve this, the sample, emitter, and detector can be enclosed in a box painted black on the inside, reducing the amount of scattering within the setup.

Optimizing these parameters is important for obtaining more accurate and reliable results. For level change, proper optimization ensures that the average readings fall within the mid-range of measurable parameters, preventing extreme readings from being clipped due to proximity to the upper or lower limits. For sensitivity enhancement, increased sensitivity allows the device to assign more precise changes in voltage readings to corresponding changes in concentration. For instance, while a less sensitive device might interpret a 0.1 V change as a 50 mg/dL difference in glucose concentration, a more sensitive device could detect changes as small as 10 mg/dL.

This level of precision is especially important when the device is intended for the exact quantification of glucose in blood. Enhanced sensitivity allows for more accurate discrimination between small changes in glucose levels, resulting in better accuracy when displaying the glucose concentration present in the body at any given time.

4 Results and Discussion

The results obtained from various tests under different conditions and optimizations provide a comprehensive understanding of the prototype's capabilities in distinguishing between increasing glucose concentrations while remaining relatively unaffected by the addition of other compounds in the sample mixture. The raw data, as presented in Appendices A-E, show the averaged results for every sample under each condition. This allows us to assess the impact of each optimization method and compare the results of un-optimized samples with those of fully optimized ones. The physical prototype is shown in Fig. 7a, alongside some of the initial un-optimized samples. The circuit schematic used to produce the prototype is illustrated in Fig. 7b, with the emitter block highlighted in red on the left and the detector block in green on the right.

The procedure for testing each sample was consistent, with minor variations. Samples were placed between the emitter and detector probes so that infrared light from the emitter would be partially absorbed before reaching the detector. A randomized selection of 10 voltage readings was taken over a 1-min period using the Arduino serial monitor. These readings were then plotted in an Excel sheet, along with their averages, for every sample under each optimization condition. The first batch of sensitivity samples is shown in Fig. 8, with labels to distinguish each sample. The measuring process, as carried out using one of the later optimized samples with red dye, is depicted in Fig. 9.

The overall results reveal a clear correlation between increasing glucose concentrations and decreasing voltage readings. On average, the un-optimized samples showed a voltage decrease of approximately 0.011 V per 50 mg/dL increase in glucose concentration, while the fully optimized samples exhibited a decrease of around 0.042 V per 50 mg/dL increase, reflecting a 280% improvement in accuracy attributable to the optimization methods. The overall correlation coefficients (R^2 values) were 0.8461 for un-optimized samples and 0.8694 for fully optimized samples. According to the correlation strength values, both are considered very high correlations, indicating that voltage readings are effective indicators of glucose concentration changes in both un-optimized and optimized tests. These results, which align with previous studies on the effectiveness of NIR absorption in glucose monitoring [12], are illustrated in Fig. 10, comparing the worst and best sensitivity test results.

In terms of selectivity, the average voltage difference between the control sample and each added compound was about 0.00095 V for individual compounds and 0.0032 V for all compounds combined in the un-optimized samples. In contrast, the fully optimized samples showed an average difference of 0.0060 V for each compound and 0.018 V for all compounds combined. As shown in Fig. 11, and

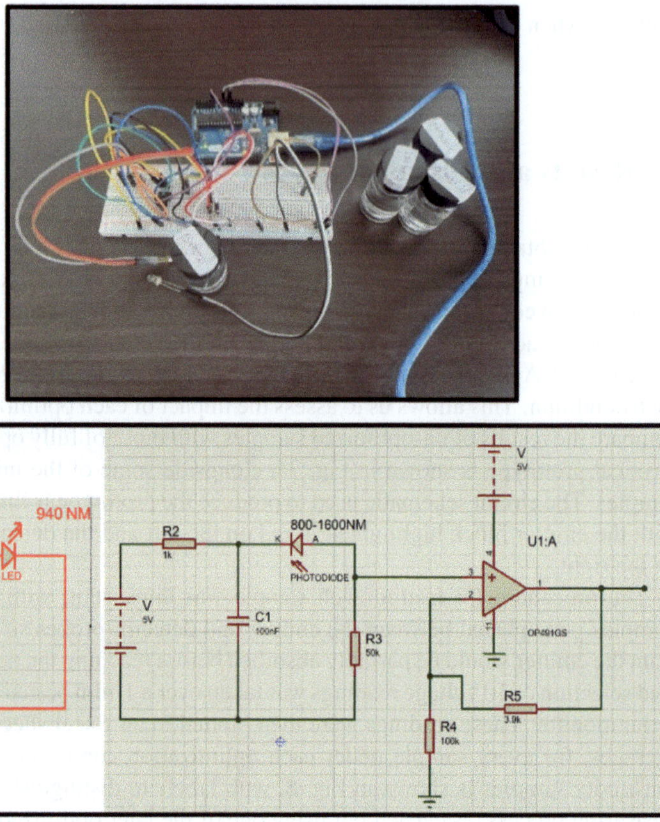

Fig. 7 **a** The fully functional prototype including some samples **b** circuit schematic diagram used to produce the prototype

especially when compared to Fig. 10, the change in voltage in the selectivity samples is much smaller than in the sensitivity samples, both in un-optimized and optimized conditions. While the un-optimized samples exhibit a smaller overall impact on glucose readings, this comes at the cost of reduced distinguishability of glucose concentrations in terms of voltage.

Beyond the results for un-optimized and fully optimized tests, the impact of individual optimization methods can be seen in Appendices B-E. These appendices demonstrate that surrounding light has minimal effect on the overall results, causing at most a 0.1 V reduction in detection levels. However, the anti-scattering cover and red dye have a much more significant impact, sometimes even overshadowing the effects of ambient light altogether. The red dye resulted in an overall reduction in detection range by approximately 1 V, while the anti-scatter cover reduced the range by about 0.4 V and improved the device's accuracy, producing a 0.3 V change in reading for every 50 mg/dL of glucose added to the sample. Additionally, there was

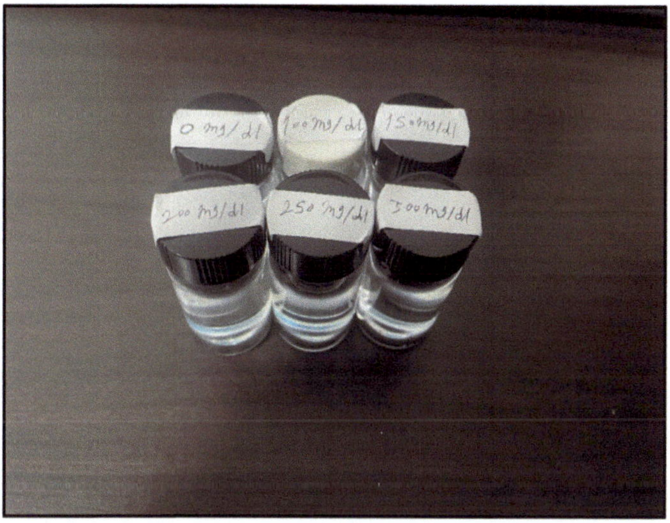

Fig. 8 First batch of samples used in un-optimized tests

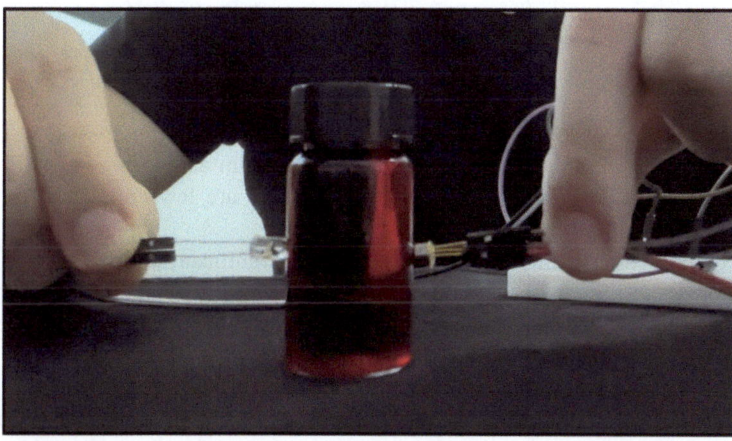

Fig. 9 The prototype in use measuring a sample with dye

a change of about 0.0061 V for each added component at its standard concentration relative to glucose.

The results from the un-optimized tests show that changes in glucose concentration, particularly in increments of 50 mg/dL, are easily detectable. The relatively high correlation factor suggests potential for this detection method to have quantification capabilities. However, it is important to note that the un-optimized tests, while promising, do not yet match the performance of similar glucose-sensing methods currently available on the market. Nonetheless, they are comparable to

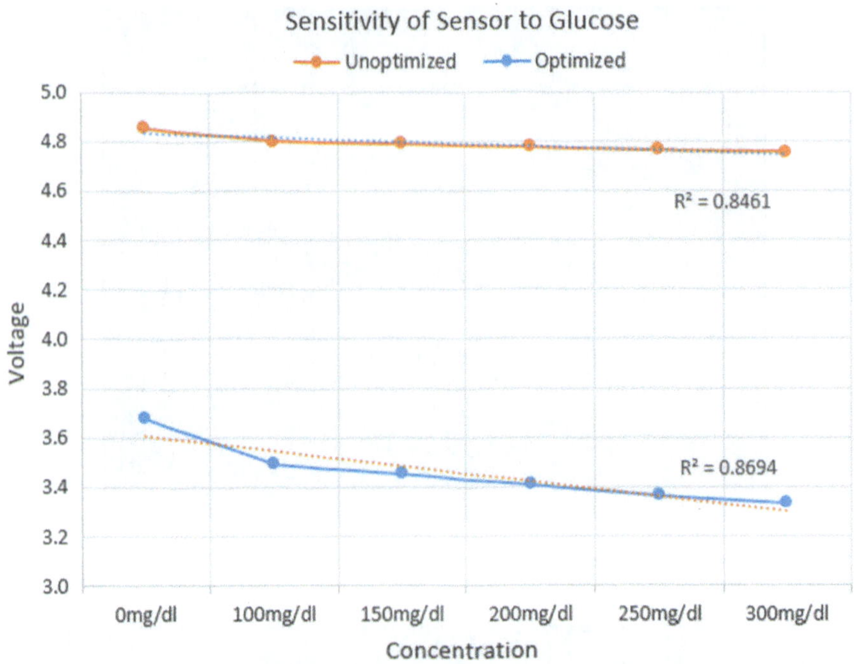

Fig. 10 Sensitivity test results comparing un-optimized and fully optimized samples

studies using different emitter ranges, such as 1100 nm and 1550 nm [3, 12–14]. Given the limited testing of this methodology, there are few comparable studies besides those mentioned.

The optimized tests, on the other hand, show a significant improvement in both correlation factors and the voltage gaps between smaller glucose concentration changes. These improvements, though simplistic, would likely exist passively in actual devices. Further enhancements, such as focusing the IR emitter, using a more suitable IR wavelength, and integrating ultrasound technology, could easily be implemented to improve results even further. An interesting observation from the optimized tests is the steep decline in voltage when comparing the sensitivity samples of 0 mg/dL and 100 mg/dL. While this decline might initially seem to result from the difference in glucose concentration, the magnitude of the change suggests other factors at play. This steep change can be more accurately attributed to the lower overall absorption of IR light by the 0 mg/dL sample, which contains only trace elements and compounds. The minimal effect of the vial holding the sample liquid also contributes to this outcome. This steep decline is likely a product of the optimizations applied, particularly the reduction of IR scattering off surfaces and into the detector, which causes higher baseline voltage readings. This optimization has also enhanced the overall sensitivity, leading to a greater difference in voltage readings between samples. The impact of surrounding lighting on the sample readings was minimal, as the detector

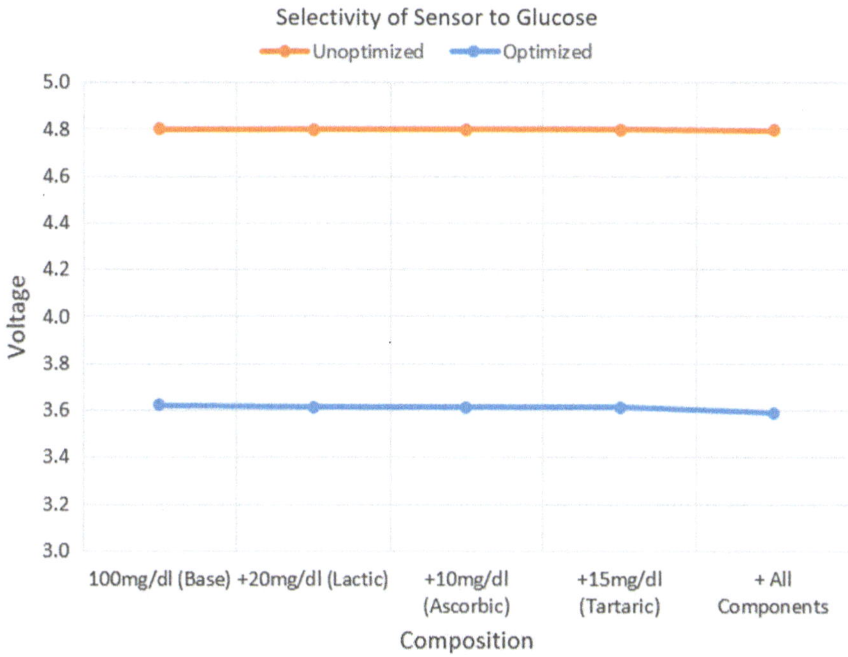

Fig. 11 Selectivity test results comparing un-optimized and fully optimized samples

was calibrated to wavelengths above the visible light spectrum, which are not typically produced by common lighting. As a result, the overall effect was a reduction in voltage readings by about 0.1 V across all samples. Conversely, the use of commercial red food colouring, which does not contain glucose, led to a decrease in voltage readings by up to 1 V. This can be attributed to the red colour aiding in IR absorption by the samples, a phenomenon somewhat comparable to the interaction of oxygenated blood with IR light.

5 Conclusion

Glucose monitoring is vital for diabetic patients, particularly for children and the elderly who may find invasive methods challenging. There's a growing need for a non-invasive alternative that is accessible and convenient for both diabetic patients and those at risk or health-conscious individuals. Near-Infrared Spectroscopy (NIR) offers a promising solution due to its skin penetration capabilities and ease of use, though it requires precise calibration and noise mitigation. This study demonstrated that by correlating voltage readings from a photodiode with glucose concentrations, NIR could effectively monitor glucose levels. Sensitivity and selectivity testing revealed a 0.011 V decrease in voltage for every 50 mg of dextrose added to a sample, with further optimization leading to a 0.042 V decrease per 50 mg, representing a 280% improvement in accuracy. Additionally, the introduction of compounds in optimized samples resulted in voltage decreases of 0.0061 V per compound and 0.018 V for all compounds combined, aligning with previous studies on NIR's potential in glucose sensing.

This study highlights the potential for Near-Infrared Spectroscopy (NIR) to become a strong candidate for non-invasive glucose monitoring, with applications in various future devices. NIR technology could help reduce prolonged hyperglycaemia in at-risk individuals by monitoring significant glucose concentration changes over time. Moreover, if integrated into common wearable technologies like smartwatches, it could provide passive glucose monitoring, offering early warnings for pre-diabetes or glucose malabsorption. With further optimization and precise signal processing, NIR devices could accurately quantify glucose concentrations, making them a valuable tool for both preventive healthcare and daily glucose management.

It is recommended that these tests be validated by third parties using more advanced technology, such as IR emitter focusing (e.g., IR lasers), and improved sample settings to reduce complications like glass vial refraction and potential contamination from chromophores. While the current results provide a solid foundation, further refinement of these preliminary tests is advised before advancing to in vivo studies to fully validate the methodology used in this research.

References

1. Ahamad Mashat, Z., Abdullah, F., Abdul Wahab, A., Faiz Md Shakhih, M., Suzziani Roslan, A.: Development of non-enzymatic screen-printed carbon electrode sensor for glucose using cyclic voltammetry. Environ. Toxicol. Manag.. **2**, 14–20 (2022). https://doi.org/10.33086/etm.v2i1.2542
2. Manasa, G., Mascarenhas, R.J., Shetti, N.P., Malode, S.J., Mishra, A., Basu, S., Aminabhavi, T.M.: Skin patchable sensor surveillance for continuous glucose monitoring. ACS Appl. Bio Mater. **5**, 945–970 (2022). https://doi.org/10.1021/acsabm.1c01289
3. Hina, A., Saadeh, W.: Noninvasive Blood Glucose Monitoring Systems Using Near-Infrared Technology—A Review. Sensors (Basel, Switzerland). 22, (2022). https://doi.org/10.3390/S22134855
4. Di Filippo, D., Sunstrum, F.N., Khan, J.U., Welsh, A.W.: Non-Invasive glucose sensing technologies and products: a comprehensive review for researchers and clinicians. Sensors. **23**, (2023). https://doi.org/10.3390/S23229130/S1
5. Van Enter, B.J., Von Hauff, E.: Challenges and perspectives in continuous glucose monitoring. Chem. Commun.Commun. **54**, 5032–5045 (2018). https://doi.org/10.1039/C8CC01678J
6. Gonzales, W.V., Mobashsher, A.T., Abbosh, A.: The Progress of Glucose Monitoring—A Review of Invasive to Minimally and Non-Invasive Techniques, Devices and Sensors. Sensors 2019, **19**, Page 800. 19, 800 (2019). https://doi.org/10.3390/S19040800
7. Mauvais-Jarvis, F.: Gender differences in glucose homeostasis and diabetes. Physiol. Behav.Behav. **187**, 20 (2018). https://doi.org/10.1016/J.PHYSBEH.2017.08.016
8. Daykin, C.A., Bro, R., Wulfert, F.: Data handling for interactive metabolomics: Tools for studying the dynamics of metabolome-macromolecule interactions. Metabolomics **8**, 52–63 (2012). https://doi.org/10.1007/S11306-011-0359-3

9. Herranz, M., Ruibal, A.: Optical imaging in breast cancer diagnosis: The next evolution. J. Oncol. (2012). https://doi.org/10.1155/2012/863747
10. Narkhede, P., Dhalwar, S., Karthikeyan, B.: NIR based non-invasive blood glucose measurement. Indian J. Sci. Technol.. **9**, (2016). https://doi.org/10.17485/IJST/2016/V9I41/98996
11. Solihin, M.I., Ang, C.K.: Blood glucose estimation using handheld near infra-red device article in international journal of recent technology and engineering (IJRTE) October. Int. J. Recent. Technol. Eng. (IJRTE). 2277–3878 (2019). https://doi.org/10.35940/ijrte.C1004.1083S19
12. Buda, R.A., Addi, M.M.: A portable non-invasive blood glucose monitoring device. IEEE Conference on Biomedical Engineering and Sciences. 964–969 (2014). https://doi.org/10.1109/IECBES.2014.7047655
13. Kassem, A., Hamad, M., Harbieh, G.G., El Moucary, C.: A Non-Invasive blood glucose monitoring device. Middle East Conference on Biomedical Engineering, MECBME. 2020-Octob, (2020). https://doi.org/10.1109/MECBME47393.2020.9265170
14. Daarani, P., Kavithamani, A.: Blood glucose level monitoring by noninvasive method using near infra red sensor. Int. J. Latest Trends Eng. Technol.. IRES, (2017). https://doi.org/10.21172/1.IRES.19

Empowering Predictive Maintenance of Medical Equipment Through AI-Driven Condition Monitoring

Aminatul Saadiah Abdul Jamil, Azira Khalil, Mardhiyati Mohd Yunus, and Jenson Gawain Fofah

Abstract This chapter explores the evolution of maintenance strategies for medical equipment in healthcare, focusing on the transition from reactive approaches to predictive maintenance (PdM) driven by Artificial Intelligence (AI). Conventional maintenance methods often resulted in inefficiencies, prolonged downtimes, and potential equipment damage. PdM, utilizing data analytics and machine learning, enables real-time monitoring and failure prediction, reducing unnecessary maintenance and enhancing operational efficiency. The study presented in this chapter employs a qualitative research design, combining a systematic literature review with the analysis of case studies from the literature and case reports from undergraduate interns and supervisors working in the Biomedical Engineering departments of four Malaysian hospitals. The research highlights the empowerment of PdM's effectiveness through AI-driven condition monitoring, demonstrating outcomes such as a 20% reduction in MRI machine downtime and significant improvements in maintenance workflows through mobile app remote monitoring. By addressing the limitations of conventional preventive maintenance, AI-driven PdM not only extends the lifespan of critical medical devices but also optimizes healthcare delivery. The chapter underscores AI's transformative potential in predictive maintenance, providing a pathway toward more efficient, reliable, and cost-effective healthcare systems.

Keywords Healthcare systems · Artificial intelligence · Machine learning · Medical device · Predictive maintenance

---

A. S. Abdul Jamil (✉) · A. Khalil
Faculty of Science and Technology, Universiti Sains Islam Malaysia, Bandar Baru Nilai, 71800 Nilai, Negeri Sembilan, Malaysia
e-mail: aminatul.abduljamil@usim.edu.my

M. Mohd Yunus
Faculty of Health Sciences, Universiti Selangor, 40000 Shah Alam, Selangor, Malaysia

J. G. Fofah
Pathology Department, Royal United Hospital Bath, Combe Park, Bath, Avon BA1 3NG, UK

K. W. Lai and P. S. Q. Yeoh (eds.), *Biomedical Engineering*, Series in BioEngineering,
https://doi.org/10.1007/978-981-97-9294-8_4

1 Introduction

The healthcare sector relies on specialized medical equipment, devices, and pharmaceuticals in laboratories, clinics, and hospitals as physical assets used to diagnose, treat, and monitor patients. The reliability of medical equipment is critical, as any failure can lead to misdiagnosis, delayed treatment, or even life-threatening situations. Moreover, medical equipment plays a vital role in preventing and controlling emerging and re-emerging diseases. To achieve optimal healthcare delivery, it is essential to maintain and regularly update the technology of medical equipment, ensuring its availability and adherence to both external and internal quality standards, thus guaranteeing the safety of patients and users [1].

As the demand for high-quality, efficient, and cost-effective healthcare continues to rise, ensuring the uninterrupted and dependable operation of medical equipment becomes increasingly vital [2]. Apart from medical equipment manufacturers, the Biomedical Engineering Department is dutybound to ensuring the installed medical equipment in healthcare facilities are in working order, since any failure to maintain its operational efficiency could lead to severe consequences, potentially compromising patient safety, equipment malfunction, and incurring significant costs to the healthcare providers [3].

The evolution of medical equipment maintenance reflects a shift from reactive approaches—characterized initially by equipment repairs or replacements undertaken post-breakdown—to more proactive strategies and data-driven maintenance [4]. The break-fix approach (also known as corrective maintenance) was long criticized for its cost-inefficiencies, prolonged downtime [3], with the potential for such maintenance to inadvertently lead to periods of idleness that were not anticipated [2], and the risk of incurring more severe damages to the equipment. At first, preventive maintenance emerged as a solution to the conundrum, involving scheduled inspections and servicing based on the original equipment manufacturer (OEM) or industry-recommended schedule to mitigate potential failures [5]. However, preventive maintenance has yet to live up to its promise, oftentimes resulting in insufficient precise failure timelines predictions, leading to avoidable machine downtime or breakdowns.

Predictive maintenance (PdM) and its latest iteration, Digital Predictive Maintenance (DPM) [6], represents a significant advancement in this area, by monitoring the medical equipment condition in situ, using data analytics and advanced algorithms to foresee equipment failures before they occur. Figure 1 shows the taxonomy of maintenance strategies used for medical equipment. The evolution towards PdM is powered by key technological advancements in Artificial Intelligence, as well as the convergence of tools such as Sensor Technology, cloud computing, IoT, and Intelligent Electronic Devices in medical equipment maintenance [7].

Artificial intelligence (AI) has made significant inroads in healthcare, by revolutionizing diagnostic, therapeutic, and operational processes. In the realm of medical equipment maintenance, AI-driven condition monitoring systems are leading the transformation toward predictive maintenance. By harnessing extensive datasets from medical devices, AI algorithms can identify patterns and anomalies indicative

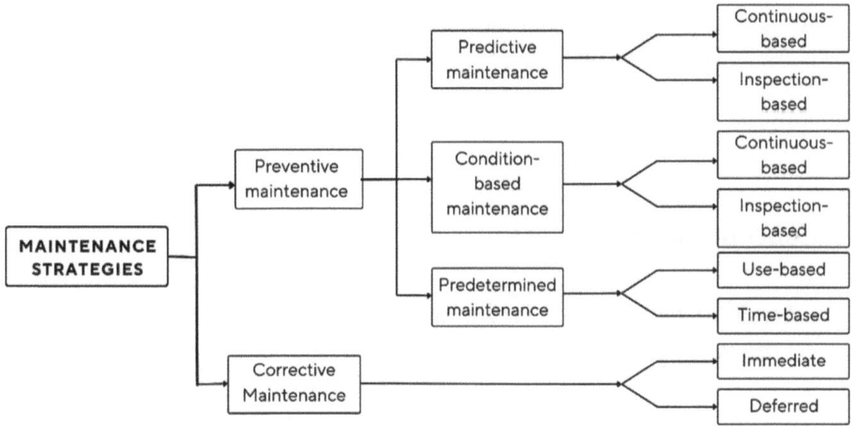

Fig. 1 Taxonomy of maintenance strategies

of potential failures, enabling timely and targeted interventions [5]. Many scholars have argued that digitization of the medical equipment maintenance and management process through artificial intelligence will reduce redundancies and introduce marked cost efficiencies [2–5].

Internet of Things (IoT), refers to a network of physical items that are equipped with sensors, software, and connections to collect and coordinate data [2]. IoT devices have become integral components of various industries, ranging from manufacturing and healthcare to transportation and smart cities. These devices continuously generate vast amounts of data through sensors that monitor environmental conditions, machine performance, and user interactions. While this influx of data offers unparalleled insights, it also poses challenges related to the management, analysis, and maintenance of these devices. Traditional maintenance approaches, often based on scheduled routines or reactive responses to failures, prove inadequate in the context of IoT devices, where the volume and velocity of data demand a more sophisticated and proactive strategy [8].

Additionally, Intelligent Electronic Devices (IEDs) have emerged as a pivotal application of technology in this context, providing advanced capabilities for monitoring, controlling, and predicting medical equipment performance. IEDs such as circuit breakers, transformers, and capacitors, serve as microprocessors and controllers within power system equipment. IEDs play a critical role in this advanced maintenance paradigm by enabling continuous data collection and analysis, which facilitates early detection of potential issues [9]. The integration of IEDs with AI technologies offers sophisticated real-time data analysis, which is instrumental in predicting equipment malfunctions and scheduling maintenance activities with precision.

This chapter explores the intersection of PdM, AI, and other integrated technologies in medical equipment within the healthcare domain. The chapter is organized as follows. Section two discusses the conceptual framework for Predictive Maintenance, AI, and related technologies in medical equipment. Section three will present selected case studies relevant to the topic, followed by a discussion of the findings, and lastly, the final section concludes the chapter.

2 Conceptual Framework

2.1 Medical Equipment Failures and Gaps in Preventive Maintenance

There is a myriad of reasons for medical equipment failures, such as failure in user handling, use of refurbished or procurement of non-original equipment or spare parts, random failure, incompetent repair technique, and wear-out failures [10]. Design defects, and other environmental factors and stress, such as electromagnetic interference, high temperatures, air quality, or humidity, may also impact the equipment performance and maintenance requirements [11].

The goal of implementing AI in PdM is to overcome the bottlenecks in routine preventive maintenance. Preventive maintenance is a time-consuming process that is prone to human error [12]. Preventive maintenance requires significant effort to manually record, manage, execute test cases, document the outcomes, and verify results. In contrast to preventive maintenance—which involves routine servicing irrespective of equipment condition—predictive maintenance is condition-based, focusing on the actual state of the equipment. This approach not only reduces unnecessary maintenance but also extends the operational life of devices by addressing issues with medical equipment only when and as they arise. When performed manually, preventive maintenance testing, which is especially essential for software maintenance, becomes tedious and counterproductive. The hidden costs associated with training new testers and managing testing environments are also often overlooked. Automation testing accelerates the process by streamlining repetitive and resource-intensive tasks.

The literature underscores inherent limitations in preventive maintenance, resulting in critical gaps that require attention. Firstly, it has been reported that on-site technicians require confirmation that the medical devices scheduled for maintenance are not in use by healthcare staff upon their arrival [13]. Additionally, the ability to easily locate these devices is crucial; however, challenges frequently arise when devices are either untraceable or have been relocated to different departments [14]. Maintenance activities were said to demand careful management to avoid work disruptions and miscommunication between healthcare providers and the technical staff. Moreover, suppliers and technicians require accurate, up-to-date information on device locations, as outdated departmental layouts can significantly impede the ability of external personnel to navigate the facilities effectively.

2.2 *Importance of Predictive Maintenance in Healthcare*

Medical devices are vital assets that must operate with high reliability and efficiency to ensure optimal patient care. Predictive maintenance plays a crucial role in this context by enabling timely servicing of equipment, thereby reducing the likelihood of unforeseen failures and ensuring continuous functionality. This approach is especially critical in hospital environments, where equipment downtime can adversely affect patient outcomes. Despite PdM could potentially be beneficial to the healthcare industry, adoption of the technology is still scant with few studies focused on the application of predictive maintenance in healthcare facilities [1, 6, 15]. While it is true that technological breakthroughs have transformed medical devices, antiquated maintenance practices are still the norm in many healthcare systems.

The practice of PdM in healthcare is currently experiencing a period of growth and expansion. PdM solutions are being implemented by more healthcare organizations to better manage their medical equipment in light of recent developments in technology and the growing prevalence of devices connected to the Internet of Things (IoT). PdM's spread throughout the healthcare industry is being fueled in no small part by an increasing emphasis on patient safety as well as the imperative to cut costs.

2.2.1 Key Components and Sensor Technologies for Condition Monitoring

Effective AI-driven predictive maintenance depends heavily on several fundamental components, including data acquisition, data processing, and predictive analytics. Vibration-based fault detection and quantification for primary flight control electromechanical actuators (EMAs) functions as fault detection methods for mechanical degradations, such as wear and localized spall, in support of predictive maintenance systems. This approach is crucial in extending the operational lifespan of EMAs, thereby reducing environmental impacts associated with frequent replacements and maintenance [9]. Vibration analysis is a commonly used method to evaluate moving parts of electromechanical systems, predicting potential failures like motor breakdowns, belt or chain slippage, and wear and tear on gears and sprockets [14].

Sensors embedded in medical devices collect real-time data on various parameters, such as temperature, vibration, and operational usage. This data is subsequently processed and analyzed using advanced AI algorithms to forecast potential failures. Medical equipment is equipped with various sensors that monitor parameters such as temperature, pressure, vibration, and electrical currents. These sensors generate continuous data streams, which are collected and analyzed in real-time. In addition to sensor data, other information sources—such as maintenance logs, equipment usage patterns, and environmental conditions—contribute to the development of predictive models [15].

2.3 AI-Driven Condition Monitoring

AI and machine learning have become central to modern predictive maintenance strategies. These technologies enable the processing of extensive datasets to uncover patterns and trends that may elude human analysts. In the context of medical equipment, AI-driven condition monitoring systems analyze data from various sources—such as sensors, historical maintenance records, and user feedback—to forecast potential equipment failures. Key technologies employed in predictive maintenance include machine learning, deep learning, i.e. artificial neural networks, recurrent neural networks [4, 16], and natural language processing. Machine learning methods can subsequently replicate different events and alterations in these twins to emphasise inadequacies and identify optimal solutions.

As shown in Fig. 2, predictive maintenance models powered by AI use historical data on equipment performance, failure trends, and environmental factors to forecast future failures. ML algorithms, for instance, can examine vibration data from a pump to find trends that indicate impending failures [14, 17]. Within AI, ML has emerged as a strong method for constructing intelligent prediction algorithms in a variety of applications. ML methods can handle high-dimensional and multivariate data, as well as identify hidden relationships within data in complex and dynamic situations. However, the performance of these applications is dependent on the suitable ML approach selection [1].

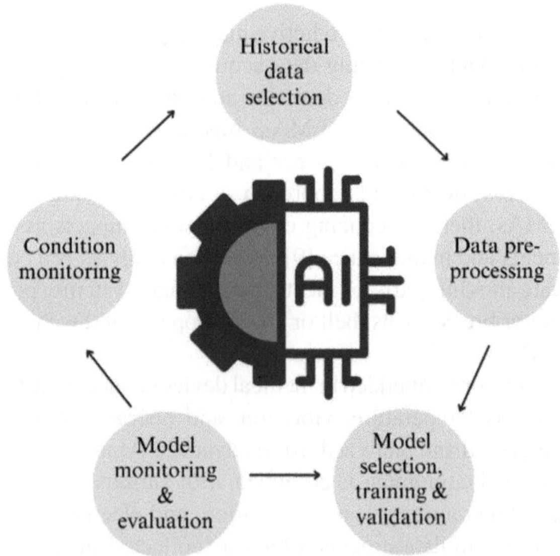

Fig. 2 AI model for a predictive maintenance application

2.3.1 Anticipatory Maintenance

Predictive maintenance, utilizing artificial intelligence (AI) to analyse real-time data from Internet of Things (IoT) sensors that monitor machine health variables, can also be advantageous for medical device manufacturing lines. These artificial intelligence models could anticipate the timing for equipment repair and send notifications to the staff-in-charge. Research indicates that implementing predictive maintenance can lead to a significant reduction in downtime, up to 50%, by proactively preventing equipment faults and minimizing the need for wasteful repair interventions [12]. Additionally, it reduces maintenance expenses and guarantees that production equipment yield a higher level of consistent quality. These enhancements lead to enhanced quality assurance for goods of medical device companies, hence reducing expensive legal ramifications.

2.3.2 Digital Twins

Medical device firms can utilise digital twins as a promising application of artificial intelligence in production. This technology use artificial intelligence to generate digital reproductions of tangible processes and medical equipment seen in the physical world [18]. Significant enhancements can arise from seemingly minor alterations. For instance, air compressors offer greater ease of maintenance and higher efficiency compared to alternative power sources for equipment, resulting in considerable improvements in uptime and energy consumption. However, manufacturers may disregard this chance because the technology that powers systems does not undergo significant changes. Digital twins could lead to solutions that humans would overlook, leading to cost savings [18]. These simulations for identifying savings can be advantageous for all manufacturers, but they hold particular significance for medical device companies. These things are frequently costly, rendering them unattainable for many individuals, but enhancements in procedures can reduce expenses and diminish clients' obstacles to participation [12].

2.3.3 Types of AI Used in Condition Monitoring

Several AI models are utilized in condition monitoring, including supervised learning models, unsupervised learning models, and reinforcement learning models. Supervised learning models are trained on labeled datasets to predict specific outcomes, such as equipment failures [11]. In contrast, unsupervised learning models identify patterns within unlabeled data, making them suitable for anomaly detection. Reinforcement learning models, which learn from interactions with their environment, are employed in complex scenarios requiring adaptive responses.

2.4 Implementation Challenges and Considerations

The implementation of AI-driven condition monitoring systems in healthcare presents several challenges, including the need for high-quality data, integration with existing infrastructure, and adherence to regulatory standards. Data privacy and security are critical considerations, given that medical equipment often handles sensitive patient information. Furthermore, the complexity of medical devices and the necessity for precise predictions demand sophisticated AI models capable of adapting to various scenarios. As for the use of sensor technology and IoT to capture data, this was suggested to be challenging, as the discrete sensor and internet coverage to all facilities incurred a significant cost in facility management [10].

3 Methodology

3.1 Literature Review and Conceptual Framework Development

The methodology of this study involved a review of existing literature on AI-driven energy management solutions in healthcare facilities. A systematic search was conducted across academic databases such as PubMed, Scopus, Google Scholar, and IEEE Xplore using keywords related to AI, predictive maintenance, and medical equipment/devices. Relevant peer-reviewed articles and conference papers published between 2020 and 2024 were identified and critically reviewed to synthesize existing knowledge and identify gaps in the literature.

The research employed a qualitative research design as it has a high level of openness which facilitated linking the investigation to existing literature, while its transparency enabled the development of a theory to account for the collected data. Thematic analysis was used to analyze the qualitative findings, and some research questions were modified and expanded based on additional insights into AI-driven predictive maintenance implementation.

The research also incorporated triangulation by analyzing 33 case reports on Planned Maintenance (Preventive and Predictive) submitted by undergraduate interns from the Healthcare Industry Technology degree program of Universiti Sains Islam Malaysia. The interns have completed a year working in the Biomedical Engineering departments across 3 public and 1 private hospitals in Klang Valley, Malaysia (Hospital Shah Alam, National Cancer Institute, Hospital Angkatan Tentera Tuanku Mizan and Pusrawi Hospital, Kuala Lumpur). Also included in the analysis are 11 supervisor modules written by the Biomedical Engineering departments' staff, where they were asked about their views among others, on the use of AI in medical equipment maintenance. Conceptually, the study was grounded in the Technology Acceptance Model (TAM). By leveraging this framework, the study investigated the perceptions of biomedical engineers and technicians regarding competency and the literacy gap in AI-drive predictive maintenance among their work force.

3.1.1 Case Studies of AI-Driven Predictive Maintenance and Empirical Validation

The research opted to use the case study approach to demonstrate the benefits of implementing AI in predictive maintenance of medical equipment, as suggested by some authors [1, 11]. The case study approach was recommended for its ability to highlight the user need for AI and what it means for predictive maintenance of medical equipment, in terms of operational efficiency and equipment reliability.

4 Results

Several healthcare facilities have experimented with machine learning (ML) for predictive maintenance in medical equipment, leading to improved operational efficiency and patient care delivery. These case studies highlight the effectiveness of ML in optimizing maintenance schedules, reducing downtime, and improving equipment reliability.

4.1 Case Study 1: Hospital A

Hospital A implemented a predictive maintenance system using ML algorithms to monitor the performance of its MRI machines. The predictive maintenance solution for MRI liquid helium pumps were based on IoT, which offers promising prospects for enhancing the availability and efficiency of these critical medical devices. The system gives a real-time data collection and recording of vibration data, coupled with AI techniques for analysis and classification. The application of IoT enables precise visibility into the condition of MRI liquid helium pumps through real-time collection of vibration data.

By analyzing sensor data and equipment usage patterns, the system predicted maintenance needs and alerted maintenance staff before failures occurred. Leveraging AI models for instance convolutional neural networks (CNN) and artificial neural networks (ANN), the system successfully predicts and classifies vibrations based on their criticality level. It enables targeted preventive maintenance and informed decision-making to prevent failures or unexpected breakdowns [14]. As a result, Hospital A reduced downtime of its MRI machines by 20% and improved the overall efficiency of its radiology department.

Once implemented, the remote monitoring of medical equipment alerted the clinical engineers of an emergent condition even when they are away from Hospital A.

Through the sub-networked real-time location system and virtual simulation system (akin to digital twin), their system provided the capability to relay alarm signals to specific tags in the location network. The AI monitoring operations was able to provide real-time reports based on pre-processed data for immediate awareness of predictive suggestions for optimal settings of equipment to meet preset ranges. Also, anomaly detection within the facility systems provides effective and efficient notifications of emergent conditions during equipment failure or impeding failure [11]. The manager or staff-in-charge was able to monitor, manipulate, and retrieve the maintenance workflow from their mobile phones. The automated system may provide awareness of current conditions, alert of potential faults, and allow the technicians to provide a certain level of repairs or system manipulation.

4.2 Case Study 2: Hospital B

Hospital B operates an inter-hospital pneumatic tube system (PTS) by connecting five different hospitals, and also run an intra-hospital pneumatic network between services on the horizontal and vertical structures on the different floors of the hospital buildings. This setup is crucial because the PTS plays a significant role in the diagnostic and treatment processes within. The PTS serves as the internal logistics means for transporting mainly blood samples, urine products, and other diagnostic materials from various departments to the laboratory.

To achieve the predictive maintenance requirement, the system responsible for the real-time display of essential installation parameters should have an alert system based on predefined thresholds. This enables timely notifications when thresholds are exceeded. Furthermore, to minimize system downtime, on one hand, the system should facilitate a reduction in Mean Time to Detection (MTTD). This can be achieved by reducing waiting time since the system will pre-indicate failures and promptly transmit notifications via WiFi to relevant technicians. The Mean Time to Repair (MTTR) is reduced as a result of proactive detection which allows technicians to anticipate the specific type of breakdown and prepare the necessary tools in advance. This optimization of time is also possible because predictive maintenance helps prevent breakdowns, leading to a significant reduction in repair durations [6].

4.3 Case Study 3: Hospital C

This case study from the literature review took place in Hospital C, where the predictive maintenance focused solely on 14 ultrasound (US) machines interspersed across different departments. Ultrasound machines were chosen because the Health Technology Manager believed they would benefit most from the proposed predictive maintenance framework, given the number of financial losses in previous years [13].

The HiWay mobile application in the case study were used by biomedical technicians in locating and maintaining medical equipment work orders without the need for rescheduling due to missing intervention. Once a technician closes a work order, he or she is forced to select a fault class among the 10 available ones to classify the maintenance. The use of the HiWay mobile application for predictive maintenance has resulted in zero lag days from an average of 41 days previously, the duration of maintenance cut by half, and the frequency of equipment downtime cut by a third.

4.4 Case Study 4: Medical Centre D

Medical Centre D implemented an anomaly detection system using ML to monitor the performance of its ultrasound machines. The system analyzed real-time sensor data to detect anomalies in equipment performance, such as unusual vibrations or temperature fluctuations, which could indicate potential maintenance needs. By proactively addressing these anomalies, Medical Centre D was able to reduce equipment downtime and improve patient scheduling.

The supervisor for Medical Centre D reported that in the event of a failure, AI not only informs the facility management personnel, but it provides awareness of the device that failed, the location, and other operational features so the responding technician brings the right equipment and tools to the job. This benefit gets the equipment back into service as quickly as possible. Through AI, the facility manager can prepare maintenance equipment supplies and tools in advance as an alternative to restoring them after a failure.

4.5 Case Study 5: Medical Center E

Siemens Healthineers developed a predictive maintenance solution for Medical Centre E's laboratory systems using Safe Automation System (SAS). The goals were to minimize downtimes, optimize maintenance, and enable long-term application. The approach involved identifying necessary sensor technology, designing a predictive maintenance framework, and automating the process. The results included proactive monitoring of system components, lower downtimes, and optimized deployment of service technicians. Siemens Healthineers aimed to make reporting malfunctions the exception rather than the rule for users of the Atellica® Solution.

The predictive service and maintenance solution aimed to significantly reduce service costs while increasing system availability. Based on the data analytics platform from SAS, the system aimed to transform clinical laboratory diagnostics to Pharma 4.0, setting new standards in immunoassay and clinical chemistry analysis. The predictive service and maintenance solution within the Guardian Program™ for Atellica® Solution is almost fully automated. Data is transmitted to country databases via the Siemens Healthineers Smart Remote Service (SRS) and then to the SAS system based on analytic models developed by Siemens Healthineers.

The five case studies demonstrate the effectiveness of AI and ML in predictive maintenance for medical equipment, highlighting its potential to improve the overall efficiency and effectiveness of healthcare delivery. By leveraging ML algorithms, healthcare facilities can proactively manage maintenance schedules, reduce downtime, and improve the reliability of critical medical equipment, ultimately leading to better patient care.

Numerous healthcare organizations have effectively adopted AI-driven predictive maintenance systems. For example, several hospitals utilize AI-powered tools to monitor MRI machines, CT scans, and ventilators. These systems have markedly reduced equipment downtime, enhanced operational efficiency, and improved patient safety. By analyzing data from multiple devices, AI-driven systems can predict potential failures with high accuracy and recommend maintenance actions before breakdowns occur.

4.6 Limitations of the Study

The study has a number of limitations. Due to time and resource constraints, the sample size for the case studies have been limited, potentially impacting the generalizability of the findings. Additionally, the study focused primarily on AI-driven predictive maintenance mainly in hospitals, and did not explore the medical equipment manufacturers. Furthermore, the study was conducted in less than a year timeframe, and long-term trends and developments in AI and predictive maintenance such as edge computing for real-time maintenance predictions have been captured in this research. These limitations were considered in the interpretation of the results and recommendations for future research.

4.7 Benefits and Challenges in AI Implementation of Predictive Maintenance

The benefits of AI-driven predictive maintenance are extensive, including reduced unexpected equipment failures, optimized maintenance schedules, and prolonged device lifespan. These systems also facilitate more effective resource allocation, leading to cost reductions and improved patient outcomes [19]. However, implementation challenges persist, such as the high costs of deployment, the need for skilled personnel to manage AI systems, and the potential for inaccuracies in predictions. One of the main challenges in implementing ML for predictive maintenance is the availability and quality of data. Healthcare facilities must ensure that data collected from medical equipment is accurate, relevant, and up-to-date to train AI and ML models effectively. Another challenge is the interpretability of AI and ML models. While AI/ML models can make accurate predictions, understanding the underlying reasons for these predictions can be challenging. Healthcare facilities must ensure that the AI/ML models are transparent and interpretable to gain trust from clinicians and maintenance staff [11].

5 Discussion

5.1 Lessons Learned from the Case Studies

The case studies highlight several key lessons. First, the success of AI-driven predictive maintenance hinges on the availability of high-quality data. Second, seamless integration with existing hospital infrastructure is crucial for effective operation. Third, continuous monitoring and updates to AI models are necessary to maintain accuracy and relevance. Lastly, collaboration between healthcare providers, equipment manufacturers, and AI developers is essential for overcoming implementation challenges.

5.2 Technological Advances in Predictive Maintenance

The field of AI-driven predictive maintenance is rapidly advancing, with emerging technologies such as deep learning offering enhanced data processing and analysis capabilities. The use of digital twins—virtual replicas of physical devices—enables more accurate simulations and predictions. The extracted predictive maintenance using digital twins were divided into the following categories: (1) Prognostics and Health Management (PHM), (2) fault diagnosis, (3) anomaly detection, and (4) RUL estimation [20]. Two cutting-edge technologies that could completely change how equipment and system maintenance are managed are PdM and DT. AI techniques like ML and DL can further the research of equipment or system performance forecasting if sufficient data from maintenance histories exists to build the most accurate model feasible. Alternatively, DT can provide a virtual representation of a physical system or piece of apparatus or system so that its performance can be monitored and its maintenance requirements can be anticipated. It enables real-time monitoring and analysis of equipment or system behaviour, allowing maintenance teams to identify issues before they become critical.

Additionally, technologies like edge computing, which facilitates real-time data processing at the source, and blockchain, which ensures data security and integrity, are transforming predictive maintenance. IEDs are also another tool at the forefront of these advancements, integrating with edge computing and other emerging technologies to provide real-time monitoring and decision-making capabilities essential for maintaining the reliability of medical equipment.

5.3 Integration of IoT, Big Data, and AI

Time Series Analysis has been widely employed in various domains for uncovering temporal patterns and trends. In predictive maintenance, TSA plays a pivotal role in understanding the temporal evolution of equipment conditions. Deep Learning for Time Series Prediction: Deep Learning, particularly recurrent neural networks (RNNs) and long short-term memory networks (LSTMs), has demonstrated remarkable capabilities in learning intricate patterns within time series data [8].

The convergence of the Internet of Things (IoT), big data, and AI is revolutionizing predictive maintenance. IoT devices generate vast volumes of data that, when analyzed with AI, offer profound insights into equipment condition and performance. Big data analytics enables the processing of these large datasets to identify trends and patterns that inform predictive models such as in the study by Zamzam et al. [15]. This integration facilitates more accurate and timely predictions, thereby reducing the risk of equipment failure and enhancing maintenance efficiency.

In the foreseeable future, the increased adoption of cloud-based platforms for data storage and processing, the development of AI-powered decision-making tools, and the creation of more sophisticated algorithms, holds the promise of managing and prolonging the lifespan of more complex medical equipment such as the robotic surgery system. The growing focus on personalized medicine is also expected to drive demand for tailored maintenance strategies that address the specific needs of individual healthcare providers.

5.4 Regulatory and Ethical Considerations in AI-Driven Maintenance

The maintenance of medical equipment is subject to stringent regulatory requirements designed to safeguard patient safety and ensure compliance with industry standards. Ethical concerns regarding patient privacy and algorithm transparency in AI adoption is also a must [21]. Regulatory bodies such as the U.S. Food and Drug Administration (FDA) and the European Medicines Agency (EMA), have established comprehensive guidelines for the maintenance and monitoring of medical devices, and Southeast Asia may be doing so in the near future. AI-driven predictive maintenance systems must adhere to these regulations to ensure that predictions are accurate and maintenance actions do not compromise patient safety.

The incorporation of AI into medical equipment maintenance introduces several ethical concerns, including potential algorithmic bias, data privacy issues, and the impact of AI-driven decisions on patient care. Addressing these concerns necessitates transparent AI development practices, regular audits, and robust data protection measures. Moreover, it is crucial to ensure that AI-driven maintenance systems complement rather than replace human expertise. Medical devices equipped with AI capabilities must adhere to ethical standards, ensuring transparency, data security, and human oversight to effectively address these ethical considerations.

6 Conclusion

AI-driven predictive maintenance represents a significant advancement in the management and maintenance of medical equipment. By leveraging AI and machine learning, healthcare organizations can predict equipment failures with greater accuracy, reduce downtime, and optimize maintenance schedules, ultimately improving patient outcomes. The integration of IoT and big data further enhances the capabilities of predictive maintenance systems, enabling real-time monitoring and precise predictions. As AI technologies continue to evolve, the potential for AI-driven predictive maintenance in healthcare is extensive. However, challenges such as the need for high-quality data, the complexity of medical devices, and regulatory and ethical considerations persist. Addressing these challenges will require collaborative efforts among healthcare providers, AI developers, and regulatory bodies. The adoption of AI-driven predictive maintenance marks a significant advancement in the healthcare industry. By equipping healthcare providers with sophisticated tools and technologies, AI-driven systems can enhance the reliability and efficiency of medical equipment, ultimately improving patient care. As the field continues to evolve, maintaining a focus on patient safety and care quality remains essential.

Acknowledgements The authors would like to thank the Faculty of Science and Technology, Universiti Sains Islam Malaysia (USIM) for supporting the current study.

References

1. Manchadi, O., Ben-Bouazza, F.E., Jioudi, B.: Predictive maintenance in healthcare system: a survey. IEEE Access **11**, 61313–61330 (2023)
2. Muhammed, A.: Distributed systems, web technology, cloud computing and IoT utilization for sustainable asset management based on AI-driven predictive maintenance in enterprise systems. J. Inform. Technol. Inform. **3**(2) (2024)
3. Kalli, V.D.R., Jonathan, E.: AI-driven energy management solutions for healthcare: optimizing medical device software. Int. J. Adv. Eng. Technol. Innov. **1**(01), 157–174 (2023)
4. Abbas, A.: AI for predictive maintenance in industrial systems. Int. J. Adv. Eng. Technol. Innov. **1**(1), 31–51 (2024)

5. Scaife, A.D.: Improve predictive maintenance through the application of artificial intelligence: a systematic review. Res. Eng. 101645 (2023)
6. Gallab, M., Ahidar, I., Zrira, N., Ngote, N.: Towards a digital predictive maintenance (DPM): healthcare case study. Procedia Comput. Sci. **232**, 3183–3194 (2024)
7. William, D., Bommu, R.: Harnessing AI and machine learning in cloud computing for enhanced healthcare IT solutions. Unique Endeavor Bus. Soc. Sci. **3**(1), 70–84 (2024)
8. Raparthy, M., Dodda, B.: Predictive maintenance in IoT devices using time series analysis and deep learning. Dandao Xuebao/J. Ballistics **35**, 01–10
9. Aderibigbe, A.O., Ohenhen, P.E., Nwaobia, N.K., Gidiagba, J.O., Ani, E.C.: Advanced sensing techniques in electro-mechanical systems: surveying the rise of smart sensors and their implications for system robustness. Eng. Sci. Technol. J. **4**(6), 323–340 (2023)
10. Abd Rahman, N.H., Hasikin, K., Abd Razak, N.A., Al-Ani, A.K., Anni, D.J.S., Mohandas, P.: Medical device failure predictions through AI-Driven analysis of multimodal maintenance records. IEEE Access **11**, 93160–93179 (2023)
11. Schmidt, M.: Predictive maintenance strategies for healthcare equipment using machine learning: utilizes machine learning algorithms to predict maintenance needs for medical equipment, reducing downtime and improving operational efficiency in healthcare facilities. Hong Kong J. AI Med. **4**(1), 56–68 (2024)
12. Khinvasara, T., Ness, S., Shankar, A.: Leveraging AI for enhanced quality assurance in medical device manufacturing. Asian J. Res. Comput. Sci. **17**(6), 13–35 (2024)
13. Luschi, A., Daino, G.L., Ghisalberti, G., Mezzatesta, V., Iadanza, E.: Empowering clinical engineering and evidence-based maintenance with IoT and indoor navigation. Future Internet **16**(8), 263 (2024)
14. Guissi, M., Alaoui, M.H.E.Y., Belarbi, L., Chaik, A.: IoT for predictive maintenance of critical medical equipment in a hospital structure. Informatyka, Automatyka, Pomiary w Gospodarce i Ochronie Środowiska **14**(2), 71–76 (2024)
15. Zamzam, A.H., Hasikin, K., Wahab, A.K.A.: Integrated failure analysis using machine learning predictive system for smart management of medical equipment maintenance. Eng. Appl. Artif. Intell. **125**, 106715 (2023)
16. Kudratillaev, M., Yakhshiboyev, R.E.: The role of indigenous medical devices and equipment development in medicine. Econ. Innov. Technol. **11**(5), 327–341 (2023)
17. Mundhra, S., Kadiri, S.K. and Tiwari, P.: Harnessing AI and machine learning in pharmaceutical quality assurance. J. Pharm. Qual. Assur. Qual. Control 19–29 (2024)
18. Mohamed, N., Al-Jaroodi, J., Jawhar, I., Kesserwan, N.: Leveraging digital twins for healthcare systems engineering. IEEE Access (2023)
19. Prabhod, K.J.: The role of artificial intelligence in reducing healthcare costs and improving operational efficiency. Quar. J. Emerg. Technol. Innov. **9**(2), 47–59 (2024)
20. Abd Wahab, N.H., Hasikin, K., Lai, K.W., Xia, K., Bei, L., Huang, K., Wu, X.: Systematic review of predictive maintenance and digital twin technologies challenges, opportunities, and best practices. PeerJ Comput. Sci. **10**, e1943 (2024)
21. Kasula, B.Y.: Framework development for artificial intelligence integration in healthcare: optimizing patient care and operational efficiency. Trans. Latest Trends IoT **6**(6), 77–83 (2023)

Multi-label Convolutional Neural Network for Multiple Diseases Detection in Chest Radiographs

Adriana Mohd Redzwan, Pauline Shan Qing Yeoh, and Khin Wee Lai

Abstract Identifying multiple diseases from a single chest radiograph is a complex and time-intensive task that requires considerable expertise. Traditional diagnostic methods rely heavily on skilled human interpretation, which can be tedious and prone to variability. Although deep learning techniques have facilitated the development of automated systems, these systems are typically limited to detecting individual pathologies. Chest radiographs, essential for diagnosing diseases such as pneumonia, lung cancer, and various abnormalities like lesions and fractures, contain substantial information that can be difficult to interpret accurately even for expert radiologists. This chapter introduces a multi-label disease diagnosis framework capable of detecting multiple pathologies concurrently. Utilizing a multi-label classification framework, this work explores two initialization strategies: transfer learning with pre-trained weights and training from scratch with random initialization. The dataset used is the NIH Chest X-ray14 dataset, comprising 112,120 chest X-rays. The results demonstrate that the Xception framework, particularly with transfer learning, outperforms the other models, achieving the highest AUC score of 0.826 and an overall test accuracy of 0.878. This study highlights the practical significance of multi-label AI algorithms in enhancing the diagnostic accuracy and efficiency of radiologists, contributing to the development of reliable and trustworthy automated diagnostic systems in the medical field.

Keywords Chest radiograph · Deep learning · Multi-label classification

1 Introduction

Recent years have witnessed remarkable progress in medical imaging technology. These advancements have not only transformed the field of diagnostic medicine but have also greatly improved the ability to detect and diagnose a wide array of medical conditions. One of the most significant breakthroughs has been the utilization of

A. Mohd Redzwan · P. S. Q. Yeoh · K. W. Lai (✉)
Department of Biomedical Engineering, Universiti Malaya, Kuala Lumpur, Malaysia
e-mail: lai.khinwee@um.edu.my

K. W. Lai and P. S. Q. Yeoh (eds.), *Biomedical Engineering*, Series in BioEngineering,
https://doi.org/10.1007/978-981-97-9294-8_5

chest radiographs. These specialized images have proven to be vital in identifying a range of chest conditions, providing crucial insights into a patient's respiratory health [1].

The global significance of chest radiographs was particularly evident during the COVID-19 pandemic. These images played a pivotal role in swiftly identifying and assessing lung-related complications, emphasizing their relevance during global health crises [2]. Additionally, the valuable information obtained from chest X-rays (CXRs), when processed through deep learning algorithms, can be seamlessly integrated into Electronic Medical Records (EMRs), offering a comprehensive patient overview for healthcare providers [3].

However, the manual analysis of radiographic images poses considerable challenges. The complex and sometimes confusing appearance of pathologies in X-ray projection images can complicate accurate interpretation. Moreover, the manual inspection of CXRs demands a significant investment of healthcare professionals' valuable time, making it a time-consuming process [4]. Interpretation inaccuracies can also arise from factors such as human error, reader inexperience, fatigue, and interruptions during analysis [5].

Patients often present with complex combinations of diseases, presenting a significant challenge for accurate diagnosis. A single X-ray scan may exhibit numerous patterns, each potentially indicative of various lung diseases. This complexity can lead to discrepancies in interpretation among radiologists, potentially resulting in unnecessary follow-up procedures [6]. Despite these advancements, there remains a critical challenge in accurately detecting multiple co-occurring conditions in chest radiographs.

In response to these challenges, there is an increasing need for automated diagnostic methods that harness the power of deep learning algorithms. While machine learning has shown promise in automating disease detection, existing models often focus on individual pathologies. This limitation necessitates a shift towards multi-label disease classification framework capable of simultaneous disease detection as depicted in Fig. 1 [7]. Such frameworks offer not only classification within a single disease but multiple findings that a CXR may associate, providing valuable guidance to radiologists in their interpretations.

Current methods for disease detection in chest radiographs lack the ability to accurately identify multiple co-occurring conditions, leading to potential misdiagnoses and treatment delays. This emphasizes the need for a comprehensive multi-label classification utilizing convolutional neural network tailored for simultaneous disease detection to improve the accuracy and efficiency of healthcare delivery. The significance of this project includes the development of multiple multi-label classification neural networks for chest-related diseases and evaluation on the impact of transfer learning in the multi-label classification frameworks.

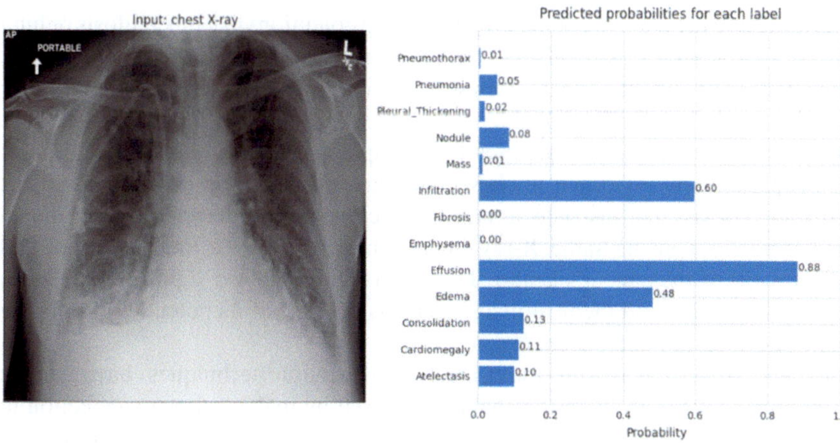

Fig. 1 Multi-label classification of chest X-ray

2 Multi-label Classification

2.1 *Single-Label Versus Multi-label Classification*

In chest X-ray image analysis, it is crucial to distinguish between multi-class and multi-label classification approaches. Both methodologies aim to diagnose diseases from medical images, but they differ significantly in their output structure and application. The differences between binary, multi-class, and multi-label classification systems are visually presented in Fig. 2.

Single-label classification (either binary or multi-class) provides the probability of the occurrence of a single disease from a medical image. This means that given a chest X-ray, the system will output a diagnosis for only one pathology, regardless of whether other diseases are present. For example, a classifier may predict pneumonia

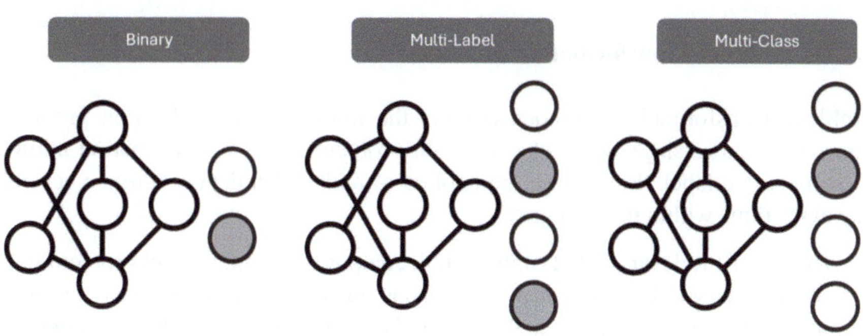

Fig. 2 Types of classification approaches

from a chest X-ray but will not identify coexisting conditions like tuberculosis or lung cancer. This approach is relatively straightforward but can be limiting in complex clinical scenarios where patients often suffer from multiple simultaneous conditions.

Multi-label classification, on the other hand, offers a more comprehensive diagnostic tool by providing probabilities for multiple diseases within the same image. This is particularly important in chest X-ray analysis, where multiple pathologies frequently coexist. For instance, a multi-label classifier can detect both pneumonia and tuberculosis in the same X-ray image, thus providing a more accurate and holistic view of the patient's health. This approach has gained significant attention in computer vision and medical image analysis because of its potential to improve diagnostic accuracy and patient outcomes [8].

Recent advancements in multi-label classification techniques have shown promising results in tackling the challenges posed by multi-pathology detection in medical imaging. Techniques such as category-wise residual attention learning and label co-occurrence learning have been developed to enhance the predictive accuracy and reliability of multi-label classifiers [9].

2.2 *Types of Multi-label Classification*

The ability to identify multiple diseases simultaneously is crucial for several reasons. Multi-label classification ensures comprehensive diagnosis by identifying all potential pathologies, thereby reducing the risk of missing a critical diagnosis. This thorough approach leads to improved patient outcomes, as it enables more accurate and complete diagnoses, guiding better treatment plans. Additionally, this method enhances efficiency by streamlining the diagnostic process, allowing for quicker and more accurate interpretations of medical images. Given the critical nature of these benefits, various multi-label learning algorithms have been developed. These can be broadly categorized into Problem Transformation methods, Problem Adaptation methods, and Ensemble methods [10].

2.2.1 Problem Transformation Methods

Problem Transformation methods convert the multi-label classification problem into one or more single-label classification problems. This transformation allows existing single-label classifiers to be applied to multi-label data. There are several key algorithms within this category:

1. The Binary Relevance (BR) approach decomposes the multi-label problem into multiple binary classification problems, one for each label. Each binary classifier predicts the relevance of a single label, and the final prediction is the combination of all binary predictions [11].

2. The Random k-Labelsets (RAkEL) creates an ensemble of classifiers, each trained on a random subset of labels (k-labelsets). The predictions of these classifiers are then aggregated to form the final multi-label prediction. RAkEL aims to capture label correlations by considering subsets of labels together [12].
3. The Classifier Chains (CC) link multiple binary classifiers in a chain, where each classifier deals with one label and takes into account the predictions of the preceding classifiers in the chain. This method captures label dependencies in a sequential manner [13].

2.2.2 Problem Adaptation Methods

Problem Adaptation methods extend traditional single-label learning algorithms to handle multi-label data directly. These methods modify existing algorithms to account for the multiple labels associated with each instance. There are several key algorithms within this category:

1. The Multi-Label k-Nearest Neighbors (ML-kNN) adapts the k-nearest neighbors algorithm for multi-label data by combining the labels of the nearest neighbors to make a prediction. This approach uses a Bayesian approach to determine the relevance of each label based on the labels of the nearest neighbors [14].
2. The MLC4.5, an adaptation of the C4.5 decision tree algorithm modifies the splitting criterion to handle multi-label data. It constructs a decision tree where each node splits the data based on a feature that maximizes the multi-label entropy gain [15].
3. The RankSVM method adapts Support Vector Machines (SVM) to handle multi-label data by optimizing a ranking loss function. RankSVM aims to rank relevant labels higher than irrelevant ones by maximizing the margin between them [16].

2.2.3 Ensemble Methods

Ensemble methods combine multiple base classifiers to improve the performance and robustness of multi-label classification. These methods can be seen as an extension of both problem transformation and problem adaptation approaches [17]. There are several key algorithms within this category:

1. The RAkEL, as mentioned is an ensemble method that creates multiple classifiers, each focusing on a random subset of labels. The final prediction is made by aggregating the predictions of all classifiers [18].
2. The Ensemble of Classifier Chains (ECC) builds an ensemble of classifier chains, each with a different random order of labels. By averaging the predictions of these chains, ECC captures more robust label dependencies and improves predictive performance [19]. The Label Space Partitioning Classifiers partition the label space into disjoint subsets and train a separate classifier for each subset. This approach helps in reducing the complexity of the multi-label classification problem and improving scalability [20].

3 Methodology

The main purpose of this book chapter is to present multi-label classification frameworks for chest X-ray diagnosis. In this work, the multi-label learning algorithm was employed using the Problem Transformation Methods, specifically focusing on the Binary Relevance (BR) approach. The BR method is a fundamental and straightforward technique for multi-label classification that transforms the multi-label problem into multiple independent binary classification problems. Each label is treated as a separate binary classification task, predicting the relevance of that label independently from the others. A separate classifier was then trained for each label to predict its presence in the given chest radiographs. This approach allows for a modular and scalable solution, where the overall multi-label prediction is a combination of individual binary predictions. The BR method's simplicity and ease of implementation make it a suitable choice for handling the diverse and complex label correlations inherent in medical imaging datasets. By employing Binary Relevance, the multi-label classification problem was effectively decomposed into manageable binary tasks, enhancing the model's ability to accurately predict multiple co-occurring conditions in chest radiographs. Figure 3 presents the overall methodology of the proposed multi-label classification framework.

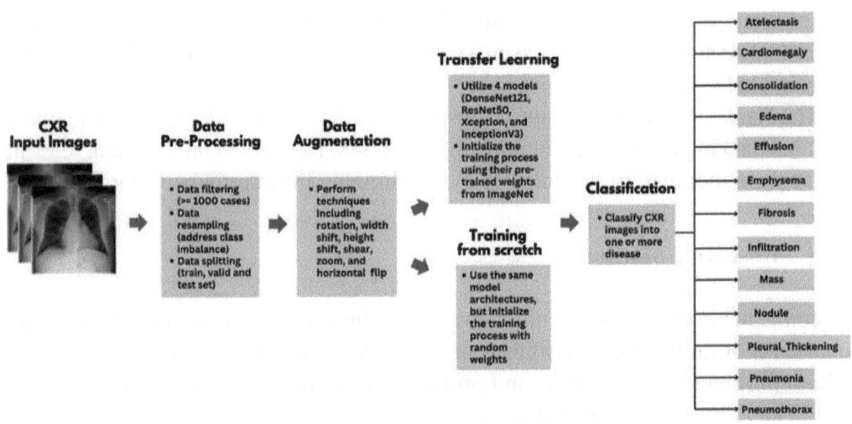

Fig. 3 Overall methodology for the multi-label classification framework

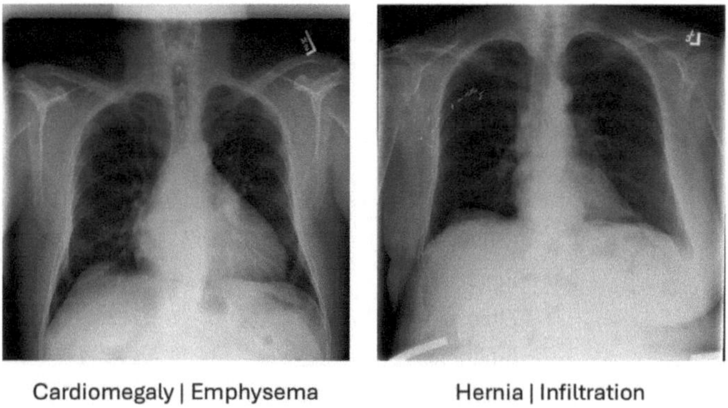

Fig. 4 Sample images from the NIH chest X-ray14 dataset

3.1 Data Collection and Preprocessing

3.1.1 Data Acquisition

The dataset used in this work was the NIH Chest X-ray14 Dataset [21]. This dataset is well-recognized within the medical imaging community due to its comprehensive collection of X-ray images, which are openly accessible to the public.

The decision to utilize this particular dataset stems from its extensive labeling that includes multiple thoracic diseases per image. This characteristic aligns with the multi-label classification focus of the study, as many of the images are annotated with more than one thorax disease. The dataset used in this work consists of 112,120 X-ray images, all standardized to a resolution of 1024 × 1024 pixels. Each image in the dataset is associated with one or more labels indicating the presence of thoracic diseases, as presented in Fig. 4. The labels were derived using natural language processing (NLP) techniques to extract disease classifications from accompanying radiological reports. The dataset encompasses 15 classes, including 14 specific diseases, which are Consolidation, Infiltration, Pneumothorax, Edema, Emphysema, Fibrosis, Effusion, Pneumonia, Pleural thickening, Cardiomegaly, Nodule, Mass, and Hernia, and one class for "No findings."

3.1.2 Data Preprocessing

Data Filtering and Balancing

To ensure the dataset includes only labels with a sufficient number of instances, labels with fewer than 1000 cases were removed. This step helps prevent overfitting on underrepresented classes and ensures the model generalizes better. After filtering, 13 disease labels were retained, as shown in Table 1.

Table 1 Number of cases per condition after data filtering

Condition	Cases
Atelectasis	11,558
Cardiomegaly	2776
Consolidation	4667
Edema	2302
Effusion	13.316
Emphysema	2516
Fibrosis	1686
Infiltration	19.891
Mass	5779
Nodule	6331
Pleural_Thickening	3384
Pneumonia	1430
Pneumothorax	5301

Resampling

To address class imbalance, the dataset was resampled to create a more balanced distribution of labels. Each sample was assigned a weight based on the number of findings it had, plus a small constant (0.04). This approach ensures that classes with fewer instances are adequately represented. The breakdown of the steps involved are as follows:

- Assigning Weights

Each image in the dataset is assigned an initial weight based on the number of disease labels it contains, plus a small constant to ensure non-zero weights. The formula for the initial weight

$$w_i = L_i + 0.04$$

where L_i is the number of labels for the i-th image

- Normalizing Weights

The initial weights are then normalized so that their sum equals 1. The normalized weight $\overline{w_i}$ for the ith image is calculated as:

$$\overline{w_i} = \frac{w_i}{\sum_{j=1}^{N} w_j}$$

This normalization ensures that the total probability across all images is 1.

- Weighted Sampling

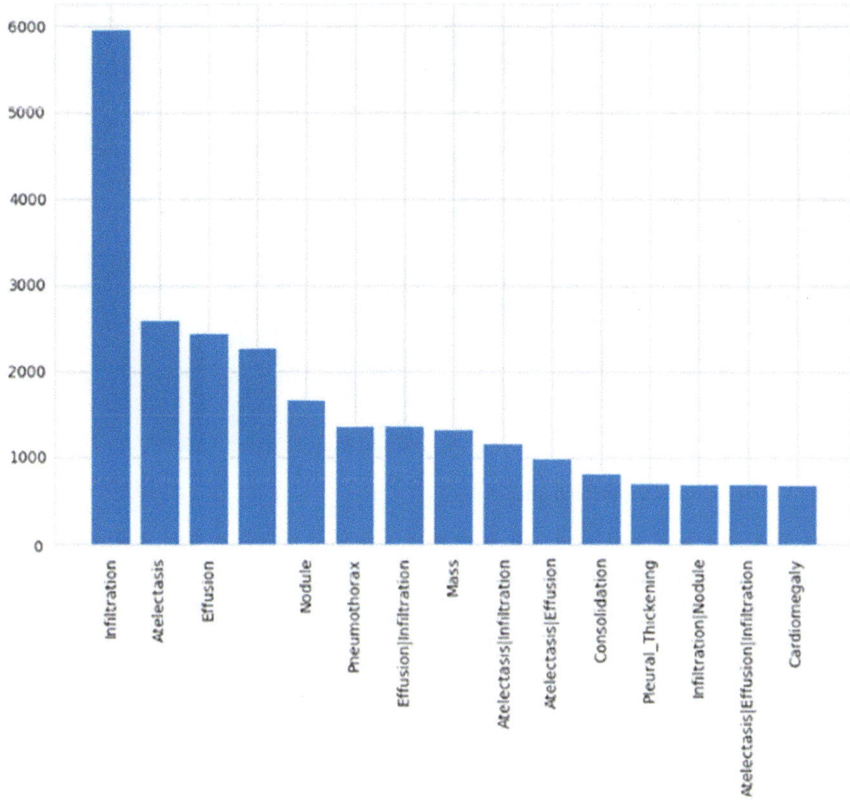

Fig. 5 Label counts

Weighted sampling with replacement was performed to create a new dataset of 40,000 images. The probability of selecting each image is proportional to its normalized weight, generated by drawing 40,000 samples according to these probabilities:

$$\tilde{D} = \{d_1, d_2, \ldots, d_{4000}\}$$

where each sample d_j is selected from the original dataset D with probability $\widetilde{w_i}$.

This resampling approach addresses the class imbalance and ensures a more reasonable distribution of disease classes in the dataset. The resulting distribution of disease classes is visualized in the bar plot as seen in Fig. 5.

Data Splitting

The dataset was split into a training set (80%) and a test set (20%). Stratification was applied based on the initial characters of the 'Finding Labels' column to ensure

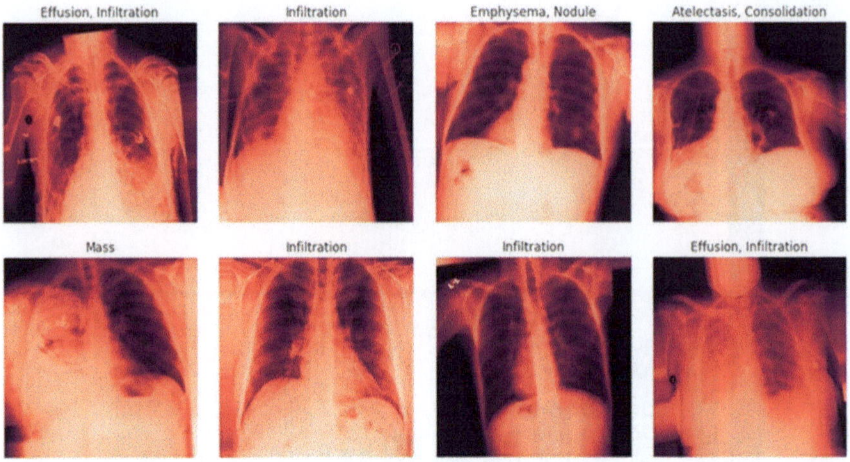

Fig. 6 Visualization of augmented images

that both sets had a similar distribution of disease labels. This helps maintain the representativeness of the dataset and ensures the model is evaluated on a realistic sample of data. The training set was further split into a smaller training set (90% of the original training set) and a validation set (10% of the original training set). This split was also stratified based on the initial characters of the 'Finding Labels' column. The validation set is used during training to monitor the model's performance on unseen data. To enhance the diversity of the training data and improve the models' generalization capabilities, data augmentation techniques such as rotation, width shift, height shift, shear, zoom, and horizontal flip were applied, as illustrated in Fig. 6.

3.2 *Model Architecture and Training*

3.2.1 Model Selection and Architecture

In this study, several pre-trained CNN architectures were selected and adapted for multi-label classification of chest X-ray images. The chosen models include DenseNet121, ResNet50, Xception, and InceptionV3. Each model was initialized with weights pre-trained on the ImageNet dataset, and a feature extraction approach was employed where the pre-trained models were used as fixed feature extractors.

DenseNet121 is renowned for its dense connectivity pattern, where each layer receives inputs from all preceding layers. This architecture ensures maximum information flow between layers, improves gradient propagation, and reduces the number of parameters. ResNet50 addresses the degradation problem in deep networks by using residual learning. The initial layers of ResNet50 were kept frozen to retain the robust feature extraction capabilities developed on ImageNet, making it suitable for adapting to new but related tasks. Xception, an extension of the Inception architecture, uses depthwise separable convolutions to enhance efficiency and performance. Lastly, InceptionV3 employs various convolutional filter sizes to capture different levels of detail in the images, thereby improving its ability to detect a wide range of features.

3.2.2 Transfer Learning and Training from Scratch

For the adaptation of these models to the task of multi-label classification of chest X-ray images, several modifications were made. The overview of the proposed training framework is illustrated in Fig. 7. Firstly, the models were configured by excluding the fully connected layers to utilize only the convolutional layers for feature extraction. Following this, a global average pooling layer was added to reduce the spatial dimensions to a single vector per channel. A new fully connected dense layer with a sigmoid activation function was added to output probabilities for each of the 13 disease classes, specifically tailored for the classification task. The models were then compiled into a complete structure by connecting the input layers to the newly added output layers, using the pre-trained networks as a backbone for feature extraction and only training the new layers. This approach effectively utilizes the transfer learning paradigm to achieve good performance with relatively minimal training on the NIH Chest X-ray 14 dataset.

In addition to transfer learning, the four models were also trained from scratch to explore the potential of learning directly from the chest X-ray dataset without pre-trained weights. This approach was employed to evaluate the models' ability to generalize purely from the available data. For models trained from scratch, the architecture remained the same as their pre-trained counterparts, but the weights were initialized randomly. The entire network was trained on the chest X-ray dataset with the same optimizer and loss function used in transfer learning. The models were

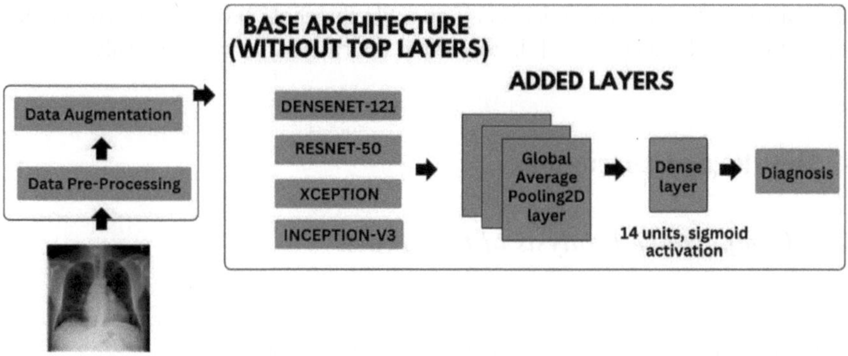

Fig. 7 Overview of proposed framework with four deep learning models

trained with learning rate of 0.001 for 20 epochs, the Adam optimizer and binary cross-entropy loss function.

3.3 Performance Evaluation Metrics

Multi-label classification is more complex than standard classification tasks where a prediction is simply right or wrong. In this classification, predictions can range from being fully correct (all labels are predicted accurately), partially correct (some labels are predicted accurately), or completely incorrect (no labels are predicted accurately). Thus, evaluating multi-label classification performance demands special metrics that consider all possible outcomes across the labels.

The evaluation metrics for multi-label classification [10] fall into two primary groups.

1. Example-Based Metrics

These metrics start by calculating the average discrepancy between the predicted and actual classes for each individual test example, then averaging these discrepancies across all test cases. Common metrics in this category include precision (PRE), sensitivity (SEN), specificity (SPE), and F1-score (F1). Precision assesses the accuracy of positive predictions, recall determines how effectively the model identifies all relevant items, and the F1-score blends both precision and recall by taking their harmonic mean.

2. Label-Based Metrics

This group includes metrics that use two averaging methods. Macro-average involves calculating a binary evaluation metric separately for each label and then averaging these metrics across all labels [22]. Micro-average, on the other hand, computes a binary evaluation metric across all labels and instances collectively. The area under the receiver operating characteristic (ROC) curve, or AUC, is particularly valued in multi-label classification for its comprehensive evaluation of a model's capability to differentiate between classes across all possible threshold levels, thus avoiding the bias of threshold selection [23]. In this work, micro-AUC is used for evaluation.

4 Results and Discussion

4.1 Comparative Analysis of the Models

Table 2 summarizes the performance of DenseNet121, ResNet50, InceptionV3, and Xception models, both with pre-trained and random weights.

The comparative analysis of various models demonstrates that utilizing pre-trained weights significantly enhances performance across different architectures.

Table 2 Overall performance of the models

Model		AUC	ACC	F1	PRE	SEN	SPE
Densenet121	Pre-trained	0.738	0.871	0.230	0.426	0.157	0.970
	Random	0.737	0.866	0.121	0.326	0.075	0.978
Resnet50	Pre-trained	0.740	0.862	0.273	0.393	0.209	0.954
	Random	0.575	0.614	0.201	0.135	0.300	0.646
InceptionV3	Pre-trained	0.770	0.878	0.144	0.576	0.082	0.991
	Random	0.753	0.876	0.101	0.508	0.056	0.992
Xception	Pre-Trained	0.826	0.878	0.378	0.512	0.391	0.960
	Random	0.749	0.876	0.073	0.504	0.040	0.995

For instance, DenseNet121 achieved a good AUC of 0.738 and an accuracy of 0.871 when using pre-trained weights, reflecting its ability to effectively distinguish between classes. However, the model showed higher precision (0.426) compared to sensitivity (0.157), suggesting a greater emphasis on correct positive predictions. Training DenseNet121 from scratch resulted in a decline in the F1-score and sensitivity, highlighting the advantage provided by pre-trained weights. For DenseNet121, the ROC curves show that the model with pre-trained weights consistently performs better across most pathologies compared to the model trained from scratch with random weights, as illustrated in Fig. 8. The pre-trained model's AUC scores range from 0.67 for Atelectasis to 0.80 for Edema. This indicates that the model is relatively good at distinguishing between positive and negative cases for these conditions. On the other hand, the random weights model exhibits lower AUC scores, with significant drops, such as Atelectasis dropping to 0.56 and Effusion to 0.66, suggesting a less reliable performance overall.

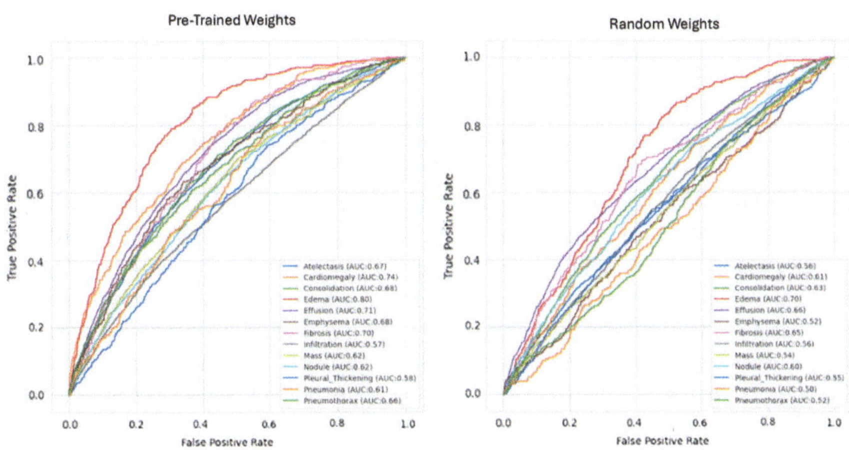

Fig. 8 ROC curves and AUC values for Densenet121

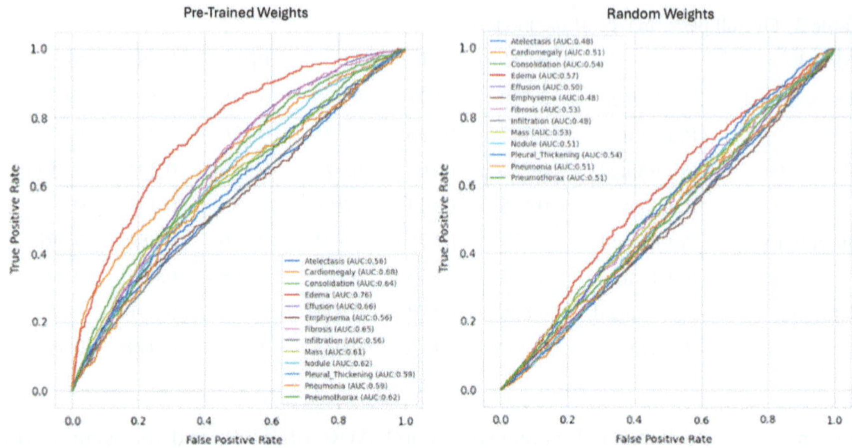

Fig. 9 ROC curves and AUC values for ResNet50

Similarly, ResNet50 exhibited robust performance with pre-trained weights, achieving an AUC of 0.740 and an accuracy of 0.862. Despite this, the model experienced a substantial drop in performance when trained without pre-trained weights, with its AUC decreasing to 0.575 and accuracy to 0.614. The ResNet50 model shows a notable difference in performance between the pre-trained and random weights configurations, underscoring the significant impact of pre-trained weights on model efficacy. As illustrated in Fig. 9, the pre-trained model achieves AUC scores ranging from 0.56 for Atelectasis to 0.76 for Edema, reflecting a solid ability to classify these conditions correctly. Conversely, the random weights model performs poorly, with AUC scores clustering around the 0.50 mark, indicating near-random guessing performance. This distinct contrast emphasizes the importance of leveraging pre-trained weights for enhanced model performance in medical image classification.

InceptionV3 also displayed commendable results, attaining an AUC of 0.770 and an accuracy of 0.878 with pre-trained weights, although its precision was notably higher than its sensitivity. Even when trained from scratch, InceptionV3 maintained strong overall performance, reinforcing the robustness of its architecture. However, it should be noted that the model achieved rather low sensitivity, with or without pretrained weights, suggesting that while the model might be good at identifying negative cases and it may fail to detect true positive cases, which could be a critical limitation in applications like medical diagnosis where missing a positive case can have serious consequences. InceptionV3, with pre-trained weights, shows strong ROC curves with AUC scores such as 0.75 (Effusion, Pneumothorax) and 0.79 (Edema) as shown in Fig. 10. The performance is notably robust across different pathologies, indicating the model's reliability. However, when trained from scratch with random weights, the model's AUC scores drop, although not as drastically as ResNet50. The AUC scores range from 0.54 (Mass) to 0.75 (Edema), suggesting that

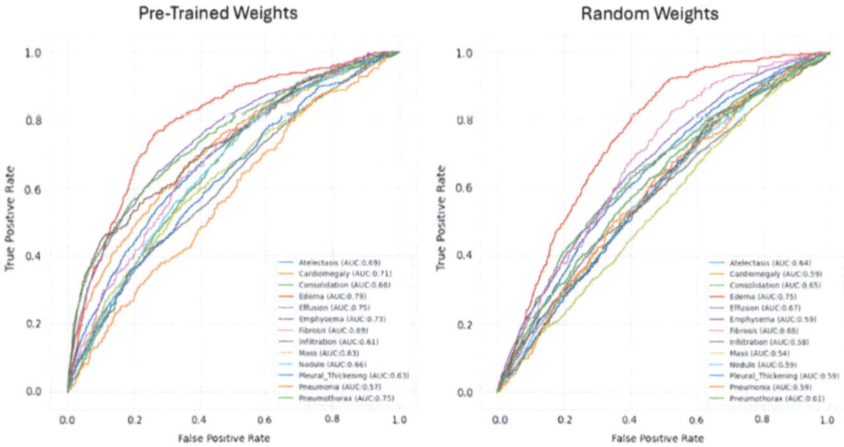

Fig. 10 ROC curves and AUC values for InceptionV3

while the model still performs reasonably well, it benefits significantly from the use of pre-trained weights.

Xception emerged as the top performer, with the highest AUC of 0.826 and a well-balanced performance across various metrics when using pre-trained weights. Even without pre-trained weights, Xception maintained a high level of performance, with an AUC of 0.749 and an accuracy of 0.876, demonstrating the robustness of the architecture. These findings emphasize the critical role of pre-trained weights in achieving superior classification accuracy and reliability in medical image analysis. The Xception model exhibits the highest AUC scores among the models evaluated, particularly when using pre-trained weights. As depicted in Fig. 11, the AUC scores for pre-trained weights range from 0.65 for Infiltration to 0.88 for Emphysema, showcasing the model's strong discriminative power. Even the random weights model shows relatively high AUC scores, with the lowest being 0.56 for Infiltration and Mass, and the highest being 0.77 for Edema. This indicates that Xception is particularly well-suited for this classification task, benefiting greatly from pre-training but also maintaining a high baseline performance when trained from scratch Among the models, Xception consistently delivered the best performance, followed by InceptionV3 and DenseNet121, while ResNet50, despite its strengths, exhibited considerable variability when not utilizing pre-trained weights. The results emphasize the importance of leveraging pre-trained weights to achieve high classification accuracy and reliability in medical image analysis.

To conclude, the ROC curves and AUC scores clearly indicate that models utilizing pre-trained weights outperform those trained from scratch across all pathologies. Among the models, Xception stands out with the highest AUC scores, followed by InceptionV3 and DenseNet121. ResNet50 shows the most significant drop in performance when trained from scratch, highlighting the critical role of pre-trained weights in achieving high classification accuracy in medical image analysis. These

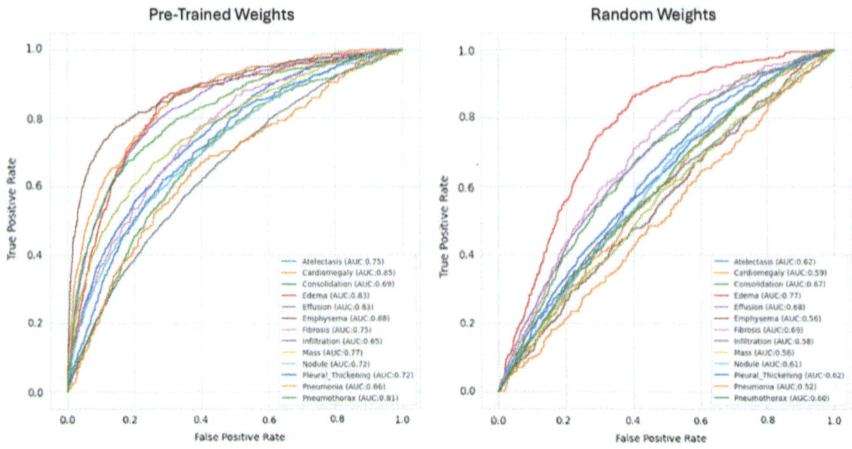

Fig. 11 ROC curves and AUC values for Xception

findings emphasize the importance of transfer learning in developing robust models for medical diagnostics. The effectiveness of transfer learning in this context proves the ability of a pretrained CNN to leverage universal features and patterns learned from ImageNet, even for vastly different domains.

However, accuracy results can sometimes be misleading, particularly when the dataset is unbalanced. Observations revealed that all models struggled to accurately predict positive cases of certain diseases, despite showing higher overall prediction accuracy and AUC values. For example, the Xception pretrained model achieved the best overall performance for test data prediction with an AUC value of 0.826 and an accuracy of 0.878. While these metrics initially suggest good performance, a deeper analysis of specific metrics such as sensitivity and precision reveals significant issues.

Table 3 illustrates that the Xception model demonstrates very low sensitivity for several critical conditions, including Consolidation, Edema, and Pneumonia, with sensitivity values as low as zero. This indicates that the model fails to identify a substantial number of actual positive cases, which is particularly problematic in a medical setting where missing a diagnosis can have severe consequences. Additionally, although the model exhibits high specificity across most conditions, this can be misleading due to the low prevalence of these conditions in the test data. Essentially, while the model is adept at confirming when a condition is not present, it struggles with accurately detecting when it is present. The PPV is also low for several conditions, suggesting that when the model predicts a condition, it is often incorrect. This is a critical concern as it can lead to false alarms or unnecessary treatments in a clinical environment.

Table 3 Test labels comparison for Xception model

Label	ACC	SEN	SPE	PRE	Prevalence
Atelectasis	0.773	0.426	0.877	0.510	0.231
Cardiomegaly	0.944	0.300	0.981	0.476	0.054
Consolidation	0.903	0.000	1.000	0.000	0.097
Edema	0.953	0.000	1.000	0.000	0.047
Effusion	0.710	0.830	0.667	0.479	0.271
Emphysema	0.957	0.292	0.993	0.692	0.052
Fibrosis	0.967	0.000	1.000	0.000	0.033
Infiltration	0.647	0.252	0.887	0.577	0.379
Mass	0.887	0.182	0.980	0.535	0.116
Nodule	0.871	0.036	0.996	0.548	0.130
Pleural thickening	0.931	0.000	1.000	0.000	0.069
Pneumonia	0.970	0.000	1.000	0.000	0.030
Pneumothorax	0.901	0.071	0.994	0.567	0.101

4.2 Limitations and Future Works

One key limitation of this study is that the proposed model was trained using only frontal X-ray images, which restricts its exposure to different views and perspectives of the thoracic region. This limitation can hinder the model's ability to capture a comprehensive understanding of various pathologies. Lateral radiographs, for instance, provide additional information that can be critical for diagnosing certain conditions. The inclusion of multiple image types could enhance the model's performance by offering a more holistic view of the thoracic anatomy, thereby improving diagnostic accuracy [24].

Another significant limitation of the proposed model is its reliance solely on the information extracted from the X-ray images, without considering the patient's medical history. Medical history, including patient symptoms, previous diagnoses, and other relevant clinical information, plays a crucial role in disease diagnosis. By incorporating such contextual information, a multi-modal model could potentially improve its accuracy and diagnostic capabilities. This integration would allow the model to make more informed decisions, similar to how a clinician considers a patient's complete medical background before arriving at a diagnosis [25].

The current approach of transfer learning in the model focuses primarily on feature extraction without fine-tuning or adjusting the latter layers of the pretrained models' architectures to better accommodate the targeted chest X-rays dataset. This approach, while effective in leveraging the general features learned from large datasets like ImageNet, may not fully optimize the model for the specific characteristics of chest X-ray images. Fine-tuning the network layers to learn features specific to the new task could significantly improve performance. Without such fine-tuning, the model might miss subtle patterns in medical imaging, thereby limiting its diagnostic potential [26].

Additionally, evaluation of this study mainly concentrated on quantitative metrics, potentially overlooking the value of qualitative assessments. Incorporating qualitative evaluations, such as visualization of affected areas [27], could provide better visual support for professional doctors through computer-aided diagnosis in clinical applications, thereby gaining their trust and helping them make rapid and accurate diagnoses. Future works can incorporate explainable AI and conduct qualitative studies of the model through implementation Gradient-weighted Class Activation Mapping (GradCAM) localization. The localization heat map of the lesion area generated by the CAM can visually explain the lesion area on which the network bases its judgments, providing valuable insights for medical professionals.

For future work, addressing the data imbalance by experimenting with oversampling the minority classes or employing undersampling techniques for the majority classes can be explored. This approach could reduce bias towards the dominant class by altering the training data to decrease the imbalance. Additionally, exploring other datasets like CheXpert [22], which provide both frontal and lateral chest X-rays, is also a potential direction. Other potential solutions to further enhance the model performance include advanced training techniques, better handling of class imbalances, or employing more sophisticated models to improve sensitivity.

While the multi-label proposed model shows promise, addressing these limitations is crucial for enhancing its effectiveness and reliability in clinical settings. Such improvements will contribute to developing more robust and accurate diagnostic tools, ultimately benefiting patient care and outcomes.

5 Conclusion

This chapter underscores the promising role of multi-label CNN models in medical diagnosis, particularly for detecting multiple chest diseases simultaneously in radiographs. By comparing the performance between models trained with pre-trained weights and those initialized with random weights, the study highlighted significant findings. The ROC curves and AUC scores indicated that pre-trained models generally outperformed those trained from scratch, proving the effectiveness of transfer learning in enhancing model performance. The findings demonstrated the clear benefits of using transfer learning in the multi-label classification framework. Pre-trained models, which utilized features learned from large datasets, showed superior performance metrics compared to models trained from scratch. The models presented in this chapter have the potential to significantly improve the healthcare system by assisting radiologists in creating more efficient radiology reports, analyzing patient disease correlations, and identifying common chest diseases. However, it remains essential to ensure that AI systems are closely aligned with human expertise, before allowing it to autonomously diagnose patients. The fusion of domain knowledge and machine learning could lead to the development of more reliable and effective diagnostic tools, paving the way for future advancements in medical diagnostics.

References

1. El-Solh, A.A., Hsiao, C.B., Goodnough, S., Serghani, J., Grant, B.J.B.: Predicting active pulmonary tuberculosis using an artificial neural network. Chest **116**(4), 968–973 (1999). https://doi.org/10.1378/chest.116.4.968
2. Albahli, S., Hassan Yar, G.N.A.: Fast and accurate detection of COVID-19 along with 14 other chest pathologies using a multi-level classification: algorithm development and validation study. J. Med. Internet Res. **23**(2) (2021). https://doi.org/10.2196/23693
3. Jabbour, S., Fouhey, D., Kazerooni, E., Wiens, J., Sjoding, M.W.: Combining chest X-rays and electronic health record (EHR) data using machine learning to diagnose acute respiratory failure. Am. Med. Inform. Assoc. **29**(6), 1060–1068 (2022). https://doi.org/10.1093/jamia/ocac030
4. Bruno, M.A., Walker, E.A., Abujudeh, H.H.: Understanding and confronting our mistakes: the epidemiology of error in radiology and strategies for error reduction. Radiographics **35**(6), 1668–1676 (2015). https://doi.org/10.1148/rg.2015150023
5. Brady, A.P.: Error and discrepancy in radiology: inevitable or avoidable? Insights Imaging **8**(1), 171–182 (2017). https://doi.org/10.1007/s13244-016-0534-1
6. Guendel, S., Ghesu, F. C., Grbic, S., Gibson, E., Georgescu, B., Maier, A., Comaniciu, D.: Multi-task learning for chest X-ray abnormality classification on noisy labels, 1–10 (2019). https://doi.org/10.48550/arXiv.1905.06362
7. Rajpurkar, P., Irvin, J., Zhu, K., Yang, B., Mehta, H., Duan, T., Ding, D., Bagul, A., Langlotz, C., Shpanskaya, K., Lungren, M.P., Ng, A.Y.: CheXNet: radiologist-level pneumonia detection on chest X-rays with deep learning, 3–9 (2017). https://doi.org/10.48550/arXiv.1711.05225
8. Guan, Q., Huang, Y.: Multi-label chest X-ray image classification via category-wise residual attention learning. Pattern Recogn. Lett. 130:259–266 (2020). https://doi.org/10.1016/J.PATREC.2018.10.027
9. Chen, Z.-M., Wei, X.-S., Wang, P., Guo, Y.: Multi-label image recognition with graph convolutional networks. In: 2019 IEEE/CVF Conference on Computer Vision and Pattern Recognition (CVPR), pp. 5172–5181. IEEE (2019)
10. Allaouzi, I., Ben Ahmed, M.: A novel approach for multi-label chest X-ray classification of common thorax diseases. IEEE Access **7**, 64279–64288 (2019). https://doi.org/10.1109/ACCESS.2019.2916849
11. Boutell, M.R., Luo, J., Shen, X., Brown, C.M.: Learning multi-label scene classification. Pattern Recogn. **37**(9), 1757–1771 (2004). https://doi.org/10.1016/J.PATCOG.2004.03.009
12. Tsoumakas, G., Vlahavas, I.: Random k-labelsets: an ensemble method for multilabel classification. In: Kok, J.N., Koronacki, J., Mantaras, R.L.D., Matwin, S., Mladenič, D., Skowron, A. (eds.) Machine Learning: ECML 2007, pp. 406–417. Springer, Berlin, Heidelberg (2007)
13. Read, J., Pfahringer, B., Holmes, G., Frank, E.: Classifier chains for multi-label classification. Mach. Learn. **85**(3), 333–359 (2011). https://doi.org/10.1007/S10994-011-5256-5/METRICS
14. Min-Ling, Z., Zhi-Hua, Z.: A k-Nearest Neighbor Based Algorithm for Multi-Label Classification, vol. 2, pp. 718–721 (2005). https://doi.org/10.1109/GRC.2005.1547385
15. Clare, A., King, R.D.: Knowledge discovery in multi-label phenotype data. In: De Raedt, L., Siebes, A. (eds.) Principles of Data Mining and Knowledge Discovery. Lecture Notes in Computer Science, pp. 42–53. Springer, Berlin, Heidelberg (2001)
16. Jiang, A., Wang, C., Zhu, Y.: Calibrated rank-SVM for multi-label image categorization. In: Proceedings of the International Joint Conference on Neural Networks, pp. 1450–1455 (2008)
17. Madjarov, G., Kocev, D., Gjorgjevikj, D., Džeroski, S.: An extensive experimental comparison of methods for multi-label learning. Pattern Recogn. **45**(9), 3084–3104 (2012). https://doi.org/10.1016/J.PATCOG.2012.03.004
18. Tsoumakas, G., Katakis, I., Vlahavas, I.: Random k-labelsets for multilabel classification. IEEE Trans. Knowl. Data Eng. **23**(7), 1079–1089 (2011). https://doi.org/10.1109/TKDE.2010.164
19. Read, J.: A pruned problem transformation method for multi-label classification. In: Proceedings of 2008 New Zealand Computer Science Research Student Conference (NZCSRS 2008), pp. 41 (2008)

20. Szymański, P., Kajdanowicz, T., Kersting, K.: How is a data-driven approach better than random choice in label space division for multi-label classification? Entropy 18(8) (2016). https://doi.org/10.3390/e18080282
21. Wang, X., Peng, Y., Lu, L., Lu, Z., Bagheri, M., Summers, R.M.: ChestX-ray8: hospital-scale chest X-ray database and benchmarks on weakly-supervised classification and localization of common thorax diseases. In: 30th IEEE Conference on Computer Vision and Pattern Recognition, CVPR 2017, pp. 3462–3471. IEEE (2017)
22. Irvin, J., Rajpurkar, P., Ko, M., Yu, Y., Ciurea-Ilcus, S., Chute, C., Marklund, H., Haghgoo, B., Ball, R., Shpanskaya, K., Seekins, J., Mong, D.A., Halabi, S.S., Sandberg, J.K., Jones, R., Larson, D.B., Langlotz, C.P., Patel, B.N., Lungren, M.P., Ng, A.Y.: CheXpert: a large chest radiograph dataset with uncertainty labels and expert comparison. In: Proceedings of the AAAI Conference on Artificial Intelligence, pp. 590–597. AAAI Press (2019)
23. Lobo, J.M., Jiménez-valverde, A., Real, R.: AUC: a misleading measure of the performance of predictive distribution models. Glob. Ecol. Biogeogr. **17**(2), 145–151 (2008). https://doi.org/10.1111/J.1466-8238.2007.00358.X
24. Albahli, S., Rauf, H.T., Algosaibi, A., Balas, V.E.: AI-driven deep CNN approach for multilabel pathology classification using chest X-rays. PeerJ Comput. Sci. **7**, 1–17 (2021). https://doi.org/10.7717/peerj-cs.495
25. Bhusal, D., Panday, S.P.: Multi-label classification of thoracic diseases using dense convolutional network on chest radiographs (2022). https://doi.org/10.48550/arXiv.2202.03583
26. Tajbakhsh, N., Shin, J.Y., Gurudu, S.R., Hurst, R.T., Kendall, C.B., Gotway, M.B., Liang, J.: Convolutional neural networks for medical image analysis: full training or fine tuning? IEEE Trans. Med. Imaging **35**(5), 1299–1312 (2016). https://doi.org/10.1109/TMI.2016.2535302
27. Baltruschat, I.M., Nickisch, H., Grass, M., Knopp, T., Saalbach, A.: Comparison of deep learning approaches for multi-label chest X-ray classification. Sci. Rep. **9**(1) (2019). https://doi.org/10.1038/s41598-019-42294-8

Artificial Intelligence in Prosthetic Gait Assessment and Prosthesis Control: Present and Future

Lai Kuan Tham, Hamza Al Kouzbary, Jia Er Kwong, Hanie Nadia Shasmin, and Noor Azuan Abu Osman

Abstract Optimal lower limb prosthesis plays a key role in rehabilitation to help restore walking ability following major limb amputations. The use of wearable sensors in prosthetic gait assessment offers a robust approach to analyze movement biomechanics of lower limb amputees in a variety of setups. The rapid development of artificial intelligence (AI) technologies has opened up new perspectives in lower limb prosthetics, from gait monitoring and outcome assessment to advanced prosthesis control and integration. With the implementation of AI algorithms, analysis of prosthetic gait parameters is readily available to clinicians, while the control of active prosthesis produces gait patterns of higher accuracies. This chapter provides an overview of various uses of AI-based methods in ambulatory gait assessment of lower limb amputees. A highlight of AI models in high-level control systems for powered prostheses and the future development of smart prostheses are also included.

Keywords Wearable sensors · Prosthetic gait · High-level control system · Powered prosthesis · Smart prosthesis · Sensor integration

1 Introduction

Human walking is a natural approach of locomotion that plays a primary role in easing short-distance navigation in day-to-day tasks and activities. Effective locomotion through precise coordination of the lower limbs assures functional versatility of the body to accommodate ambulation on different terrains (e.g., level ground, stairs, slopes, uneven surfaces, etc.) and obstacles (e.g., curb, rocks, drain, etc.) [1]. As such, preserving normal gait has always been the main objective in human locomotion due to its optimal physiological performance [2]. Limb loss due to trauma or medical conditions such as peripheral vascular disease, cancer, diabetes, infections and congenital defects [3, 4] leads to mobility impairments that bring adverse

L. K. Tham (✉) · H. Al Kouzbary · J. E. Kwong · H. N. Shasmin · N. A. Abu Osman
Centre for Applied Biomechanics, Department of Biomedical Engineering, Faculty of Engineering, Universiti Malaya, 50603 Kuala Lumpur, Malaysia
e-mail: laikuan.tham@um.edu.my

K. W. Lai and P. S. Q. Yeoh (eds.), *Biomedical Engineering*, Series in BioEngineering,
https://doi.org/10.1007/978-981-97-9294-8_6

impact to normal gait. As one of the major contributors to major lower limb amputations, diabetes recorded a prevalence of 10.5% or approximately 537 million people worldwide in 2021, with a projection to 12.2% or 783.2 million by 2045 [5]. This is in accordance with the rising number of major lower limb amputations performed globally each year [6].

The gait of lower limb amputees is typically different from able-bodied gait due to the loss of major muscle groups following lower limb amputation [7–10]. Prosthetic gait can also be affected by numerous factors, such as prosthetic alignment, socket fit, and the adequacy of prosthetic components [11]. Understanding the gait characteristics of lower limb amputees is thus essential in clinical rehabilitation to help restore mobility and quality of life. Gait assessment is the common clinical approach used to evaluate prosthetic gait as an assisting tool in lower limb prosthesis prescriptions [12]. In terms of prosthesis design, gait analysis provides supportive information in the functional evaluation of prosthetic components. The approach is also important in monitoring rehabilitation and therapy outcomes, evaluation of mobility efficiency, and fall risk assessment [13]. In this regard, assessors with adequate knowledge of human gait biomechanics, both normal and pathological, are highly demanded to identify gait deviations and the underlying skeletomuscular mechanisms which affect the gait quality of lower limb amputees [14, 15].

The gait of individuals with lower limb amputation is often evaluated through observational gait analysis, where human eyes are frequently the most accessible tool in the clinical environment [16, 17]. Observational gait analysis is normally performed with functional tests such as the Timed Up and Go (TUG) Test, Six-Minute Walk Test (6 MWT), and 10-Meter Walk Test (10 MWT), by observing the gait pattern, joint and segmental movements of the body from different viewing angles. Abnormalities observed will be compared with normal gait or mapped with the typical patterns of deviation to classify gait performance which helps in rehabilitation strategies and lower limb prosthesis refinement [2, 18, 19]. The approach is generally limited by its subjective outcome in which consistency could be greatly affected by the variation in knowledge, skills, and experience across the assessors. Accuracy is further affected by the high speed and complex movements during locomotion which are extremely difficult to be captured through visualization.

Instrumented gait analysis offers an objective means for prosthetic gait assessments. The approach, commonly recognized as the gold standard, involves motion capture devices and force platforms to conduct gait assessments under a controlled environment. Comprehensive and accurate analysis of gait and joint movements has become possible with quantitative measures provided by the system. Nevertheless, the setup of the system is complicated and often associated to high. Skilled personnel are also required to acquire gait data in a dedicated working space [20, 21]. Thus, the application of the camera-based system is generally very limited in the clinical settings.

Significant development of wearable technology in recent years has opened up new possibilities for prosthetic gait assessment. Wearable systems comprise miniature sensors offer a blend of convenience and robustness, the portable sensors al-low gait measurements in various environments, including outdoor and real-life activities.

Such systems can be used in long-term gait monitoring, in which data collected can be easily integrated with the health monitoring systems to obtain continuous feedback from the end users. As a revolutionary tool in gait analysis, artificial intelligence (AI) offers sophisticated approaches for processing complex features embedded within the gait patterns of human locomotion. This chapter provides insights into the application of AI in the gait assessment of lower limb amputees using wearable sensors, and the advancement of smart active prostheses control that helps improve mobility of lower limb prosthesis users.

2 AI Techniques in Prosthetic Gait Characterization

Gait characterization and quantification are essential in clinical applications for providing objective feedback during the planning phase and outcome evaluation of clinical interventions throughout the rehabilitation procedures [2, 20, 22]. Furthermore, characterizing gait patterns and events plays a major role in optimizing the design and functionality of lower limb prostheses [23, 24]. Data collection of prosthetic gait has become more convenient thanks to the rapid advancement of sensor technology, as wearable sensors permit measurements in various conditions. Nevertheless, translating signals collected using wearable sensors into physiologically meaningful measures remains a significant challenge due to the complexities of human locomotion. AI-based methods have become a popular domain in gait parameters computation following its remarkable evolution in the past decades [25]. The integration of AI in analyzing prosthetic gait has shown great potential in improving the accuracy of output measure from inertia measurement units (IMUs), leading to significant advancements in gait assessment, optimization, and prosthetic design for lower limb amputees [24, 26].

2.1 Gait Phases and Events

A complete gait cycle consists of two phases: (1) stance in which the foot is in contact with the ground and (2) swing in which the foot is raised from the ground. Events of a gait cycle, i.e., initial contact, foot flat, heel off, and toe-off, are temporal sequences performed by the foot as the body advances during locomotion [2, 19]. Analyzing the timing of gait events such as stride time, single-, and double-limb support time allows an understanding of symmetry and coordination between the lower limbs within the cyclical gait pattern. Besides, gait phases or events are particularly useful in providing input signals to generate precise functionality control for active prostheses [25].

Various methods of gait phase or event detection methods based on AI were developed. For instance, the Hidden Markov Models were used [27] to detect gait events from the sagittal foot angular velocity measured by a single IMU. Accuracies of the proposed machine learning model were reported to be higher than the threshold-based algorithm that classified gait based on peak detection accelerometer and gyroscope signals.

In terms of neural networks, Jung et al. [28] implemented a multilayer perceptron neural network (MLPNN) to detect the stance and swing phases for the control exoskeleton ROBIN-H1. The study reported an average classification success rate of 97.75% for the offline test and 90.75% for the online test. Similar MLPNN was implemented [29] to classify the stance and swing phases based on electromyography signals, with accuracies ranging between 79 and 87% for different optimization methods. [30] detected four gait events using an MLPNN and reported average classification success rate of 83.17, 82.42, and 83.23% for 3, 5, and tenfold cross-validation, respectively. The deep learning models have also shown outstanding performance in gait event detection and classification. A long short term memory algorithm and deep neural network was applied to classify gait phases using the accelerometers, which recorded accuracies >91.8% in one study [31].

The Support Vector Machines (SVM) used demonstrated accuracies of 83.00, 82.69, and 83.29% for 3, 5, and tenfold cross-validation, respectively, in the detection of four gait events [32]. On the other hand, an SVM model developed in a study [33] exhibited the highest accuracies among all proposed neural networks in the study.

2.2 Gait Kinematics

Prosthetic gait kinematics, e.g., body segment orientation, joint angles, velocities, and accelerations, generally indicate bilateral movements of the lower limb segments during ambulation. With the aim to overcome drift errors originated from the inertia sensors, a nonlinear autoregressive neural network with exogenous inputs (NARX) was proposed to estimate 3-dimensional (3D) body segment orientation for lower limb amputees. Results of the study showed the ability of the proposed neural network in replicating the robustness of a complementary filter, while avoiding drifts produced by the accelerometers of the IMUs. The proposed model also demonstrated the dynamic properties of estimating segmental orientation in various types of locomotion tasks [34].

Machine learning algorithms using dynamic time warping (DTW) and Euclidean distance in combination with the k-nearest neighbors were shown to be feasible in classifying prosthetic gait using wearable gyroscopes. The proposed models produced gait profiles which captured changes in gait patterns could be useful in performance assessments of lower limb amputees following gait training [35]. Gait abnormalities can also be classified in real-time using a machine learning model based on Extreme Learning Machines [36]. Furthermore, a long short-term memory (LSTM) deep learning model has shown promising accuracy in the fall risk classification of lower limb amputees during the 6 and 2-min walk tests [35].

In addition, prosthetic gait kinematics is frequently analyzed as feedback for the optimization of alignment, stiffness, socket fit, and other mechanical properties of powered lower limb prostheses [36], which will be discussed in the following sections.

3 AI in Powered Lower Limb Prostheses

As aforementioned normal gait provides the optimal physiological performance. Therefore, recouping the normal gait is essential for amputees to reintegrate into society. The most commercially available prostheses are the passive prostheses which are known for asymmetric gait, low shock absorption, and high metabolic cost [37, 38]. Passive prostheses might lead to joint disorders, chronic residual limb pain, back pain [39], and increase the risk of developing osteoarthritis [40–42]. Due to poor mobility and reduced use of prostheses shortly after amputation, some patients have depression or depressive symptoms [43]. Patients who underwent long-term amputations also attributed their depression to activity restrictions, overwhelming feelings of vulnerability, and poor self-rated health [43–45]. Powered prosthetic devices improve the mobility and quality of life of amputees, as they mimic the biological limb behavior, providing a more symmetric gait pattern and net positive power during the stance phase, which also reduces the metabolic cost [46]. Moreover, amputees using powered prostheses achieve higher clinical scores in gait speed, balance and safety compared to passive prostheses users [47]. Despite the advantages of powered prostheses, these systems are still limited and not commonly used by patients. The high-cost and some performance limitations are the main factors that limit the wide spread of robotic prostheses [23, 48]. Thus, ongoing research is still improving powered prostheses, both mechanically and in terms of control.

Based on the control scheme of powered prostheses we can categorize the controllers into seven different categories: (1) Echo controllers [49–51]; (2) Two-level finite state machine (FSM) controllers [52–64]; (3) Three-level conventional controllers [65–72]; (4) Virtual constraints and inverse dynamic based controllers [73, 74]; (5) Proportional Electromyography (EMG) volitional continuous controllers [75, 76]; (6) Biomimetic neuromuscular model-based controllers [77, 78]; and (7) Reference pattern generator-based controllers [79–82].

The development of powered prostheses begins in the mid-sixties [83]. However, it did not gain a lot of attention until the first decade of this century. AI has an important role in the development of powered prostheses as most of the existing controllers have machine learning or deep learning algorithms embedded in the system. Regression, classification models, fuzzy systems, supervised and unsupervised learning are used to control the powered prostheses. The AI algorithms used in the controllers are presented in the rest of this section.

3.1 Echo Controllers

Four microswitches at the sole of both feet were used to detect the walking phases [84]. During the swing phase, the knee angle of the sound leg is fed after a delay function to a fuzzy controller to control the knee, while during the stance phase the knee joint is locked.

A fuzzy inference system was developed in [85] to detect the gait phase and generate a nominal output torque and desired knee angle corresponding to the determined phase based on the angle of the femur and the shank of the intact limb.

3.2 FSM Two-Level Controllers

Usually, the FSM two-level controller relies on threshold predetermined values to switch between phases and the impedance, proportional-derivative, or torque controllers' parameters are constant for each phase. However, many studies have introduced AI algorithms to switch between the walking phase or to provide adaptive updates to the controller parameters.

In [86], the gait cycle was divided into five states, and two algorithms were developed and compared for gait phase detection. First, the authors presented a rule-based estimation algorithm, consisting of four threshold equations. Second, they introduced an artificial neural network (ANN) algorithm. The inertial measurement from the shank of the prosthesis (i.e., gyroscope data) was selected to be the input data for the ANN and the image-based measurement was defined as the target data. The results show that the ANN had slightly better phase-detection for both low and high walking speeds.

Authors in [52–56, 58–64, 87] have implemented several AI algorithms to provide an adaptive update to impedance control variables. In [58], the FSM consists of five states, i.e., initial double support, single support, terminal double limb support, swing flexion, and swing extension. An auto-tuned adaptive impedance controller was developed, a fuzzy controller and the mean of maxima defuzzification method were implemented to decrease or increase the stiffness and equilibrium position coefficients of the impedance controller. A Dempster-Shafer theory-based transition rule was implemented by authors in [56] instead of the threshold transition rules to switch from initial double support to signal support and from signal support to terminal double limb support. In [54], a fuzzy logic tuner was implemented to tune the impedance control parameters based on human expert tuning. The inputs of the fuzzy tuner were the difference between the prosthesis and target gait peak knee angle, gait phase duration, and peak angular velocity. 24 rules dictate the setting of the values of the impedance controller's parameters for each phase of the five phases. Adaptive Dynamic Programming (ADP) was used to tune the impedance controller parameters (Fig. 1). The authors applied and compared the results of two algorithms, namely, direct heuristic dynamic programming (dHDP) and neural fitted

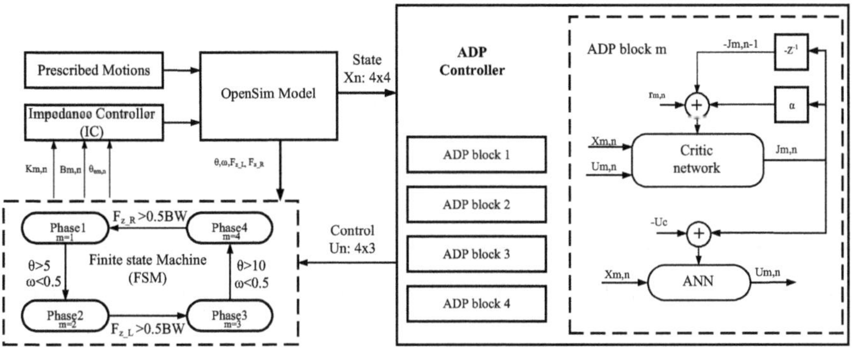

Fig. 1 This diagram shows the ADP algorithm used for updating the impedance controller parameters

Q with continuous action (NFQCA). For each of the FSM phases (four phases instead of five), there is a corresponding dHDP/ NFQCA block. The output of the action neural network is used to update the impedance control parameters. Authors in [59] compared the efficiency of parallel tuning where control parameters are all tuned during the stance and the swing phases at the same time, and sequential tuning where control parameters are tuned during the stance or swing phase while fixing the remaining parameters. In [55], developed another ADP algorithm based on offline policy iteration reinforcement learning to configure the optimal parameters of the impedance controller by utilizing historical data. In [64], a reinforcement learning tracking control scheme based on dHDP was developed. Authors in [53] used a flexible policy iteration (FPI) with historical data instead of their dHDP module. Authors have compared the FPI method with other reinforcement learning-based methods such as dHDP, NFQCA, and generalized policy iteration (GPI). A user-controlled interface (UCI) combined with RL based controller was used by other author in [52], the UCI allowed to define the desired prosthesis knee impedance. A policy iteration with constraint embedded (PICE) method was proposed in [87] to reduce the tuning time.

A fuzzy system was used to estimate the friction torque during the swing phase in [88]. The impedance controller parameters are changed based on the state of the prosthesis. Then, the output torque will be added to the estimated frictional torque.

3.3 Conventional Three-Level Controllers

In these controllers, AI plays an essential role at the high level to recognize the walking task. These controllers consist of three levels: At the high level the intent recognizer and at the mid-level variant FSMs each FSM represents the phases of a waking task. Finally, the low-live could be variant controllers e.g., torque, position, and impedance, or it could be an impedance controller with variant parameters

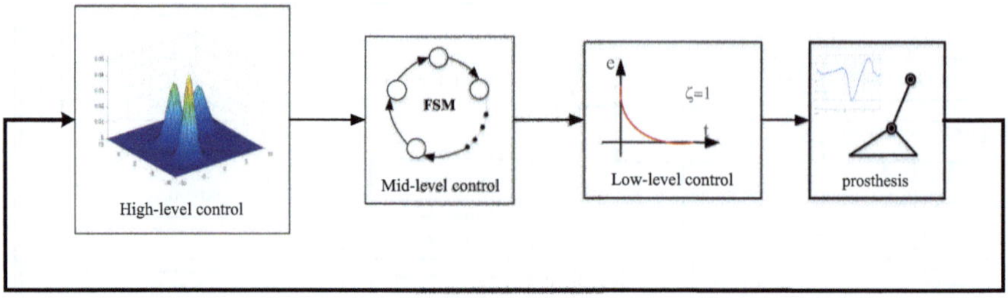

Fig. 2 Block diagram depicts the hierarchical three-levels control strategy used to control most powered prosthetic devices

(change with the walking phase). Figure 2 shows the general structure of conventional three-level controllers.

3.4 Proportional EMG Volitional Continuous Controllers

EMG volitional continuous controllers work on mapping the relationship between the residual muscles' activation and the knee or ankle movements.

Quadratic discriminant analysis (QDA) was used to identify the flexion–extension movement and a principal component analysis (PCA) was used to estimate the desired reference velocity fed to the impedance controller [89, 90]. Subjects were asked to track four different trajectories while sitting on a chair. In trajectory A, the subjects were asked to change their knee angle quickly from 8° to 45° increments and hold for 5–10 s. Trajectory B were sloped trajectories to test the subjects' capability of moving the prosthesis at different constant speeds. Finally, trajectories C and D were sinusoidal waves at 0.2 and 0.33 Hz, respectively. The average root mean square error (RMSE) for all subjects for the four different trajectories A, B, C and D were 7.7° ± 076, 3.4° ± 0.76, 5.6° ± 1.4 and 7.9° ± 0.72, respectively.

Linear Discriminant Analysis (LDA) pattern recognition to identify medial rotation and lateral rotation of the foot, ankle dorsiflexion and plantarflexion, ankle eversion and inversion, and a relaxed class was used [91]. The highest accuracy for both able-bodied subjects and transtibial subjects led to 0.2 ± 0.1 and 4.7 ± 0.5% classification error, respectively.

In [92], an LDA classification algorithm has been implemented to recognize the ankle movement for both the amputated and the sound limbs. Subjects were firstly instructed to perform five movements of dorsiflexion and plantarflexion and they were asked to pause in between. Secondly, they were asked to perform the movements continuously without pause. The classification accuracy of the ankle movement was 100% for the tibialis anterior and medial gastrocnemius, and it was 99.923% for the lateralis gastrocnemius.

Three classification algorithms were implemented and compared (i.e., QDA, SVM, and k-nearest neighbors (kNN)) [76]. The SVM algorithm had the best performance compared to QDA and kNN. The average classification accuracy for the four movements was 91.85 ± 3.57%.

4 Reference Pattern Generator-Based Controllers

A central pattern generator (CPG) based on the Rayleigh model was implemented in [93, 94]. Authors tuned the CPG parameters based on the height, weight, and age of each amputee using a fuzzy logic classifier to generate rhythmic patterns for level ground walking (LG).

In [95–99], the control system consists of three parts. First, the feature extraction is done through a recursive mean squared (RLS) technique. Second, an unsupervised Kohonen self-organizing map (SOM) neural network is used for dimension reduction to reduce the noise and absorb the large variations in the original features. Therefore, making the supervisory network in the next step converge faster. Finally, a feed-forward neural network to generate the knee angle during LG. Authors have improved their method by adding extra input data which is proprioceptive data from a gyroscope sensor attached to the residual limb in [96]. While authors in [98] edited the control structure, the proprioceptive data is used to predict the knee angle and the predicted angle is used to correct the estimated data done by the neural network. Authors developed an extended Kalman filter (EKF) to estimate the foot orientation with respect to ground using two gyroscopes and four infrared sensors mounted on the foot sole in [99]. The error to signal percentage for the first subject were 5.12 ± 0.99 and 5.73 ± 1.06, respectively. While for the second subject, the errors were 5.21 ± 1.44 and 5.96 ± 1.69, respectively [96, 98].

A NARX was developed to estimate prosthesis ankle angle during LG using the EMG data measured from the residual limb's muscles using embedded electrodes within the socket and transtibial prosthesis's kinematic data as a feedback signal to the neural network [100, 101]. The average confidence level of the neural network for the three subjects was 91% and the average RMSE was 3.07°.

Authors in [102] extracted EMG features in addition to the IMUs features from the shank of the sound leg and the thigh of the affected side, and the force sensitive resistors data on the heel and the forefoot are all fed to a two-layer feedforward ANN to identify the walking states (standing, walking, starting, and stopping). Also, another two-layer feedforward ANN was used to predict the percent of gait cycle (PGC). The average RMSE for LG and ramp ascend (RA) was 4.43°.

A Gaussian process (GP) regression was trained in an attempt to estimate the motor position of the prosthetic/orthotic foot [103, 104]. This algorithm relies on the data of the shank's angle and angular velocity measured by a gyroscope attached to the shank. A similar GP algorithm was used to predict the reference knee angle [105, 106]. A support vector regression (SVR) model was implemented to predict the ankle angle and moment using the knee and hip joints' kinematic data as input [107].

A random forest regression model has been developed by authors to estimate the ankle and knee angles and moments using the thigh kinematic data [108, 109]. The authors have trained the random forest model based on two input data sets. First, the thigh orientation and angular velocity. Second, they add thigh acceleration to the first set. The GP had the best performance when it was trained with four walking speeds. The average RMSE, the maximum absolute error, and the coefficient of correlation (R^2) for the five walking speeds between the desired and estimated terrains were 1.5 e^{-3}, 3.32 e^{-3} m, and 0.95, respectively. In [105, 106], the GP failed to produce a proper knee trajectory using a single input signal thigh angular speed or thigh angular position. However, when using both signals the GP was able to generate the required knee angle trajectory for different walking speeds. The GP had the best performance when it was trained on the first subject data for all walking speeds. The GP was tested based on the second subject data to have an average RMSE for all walking speeds of 6.36°, and the average maximum absolute error for all walking speeds was 14.04°. While the coefficient of determination of the five walking speeds was 0.894. When the knee and hip angles in addition to the knee and hip angular velocity were used as inputs, the SVR had the best performance with average RMSE of 1.29° and $0.08\ \mathrm{N \cdot m \cdot skg^{-1}}$ [107]. The second data set had the best performance with average RMSE of ankle angular position, moment and knee angular position, moment 1.804°, 0.119 N/kg, 1.216°, 0.057 N/kg, respectively [108, 109]. The correlation between the actual and estimated ankle angular position, moment, and knee angular position and moment were 0.966, 0.941, 0.995, 0.951, respectively.

In [110], the controller receives the input data of 10 selected muscles' EMG signals of the lower limbs, and a deep belief networks (DBN) that consists of restricted Boltzmann machines (RBM) is used to obtain the optimal feature vectors. Finally, a feedforward neural network generates the desired trajectories for the lower limbs' joints. The generated trajectories were accurate for all joints as the average RMSEs for all walking speeds were $3.58° \pm 0.67$, 3.96 ± 0.69, and 2.45 ± 0.57 for the hip, knee, and ankle, respectively.

The method proposed in [111] consists of many stages including a comparison of two prediction models (i.e., Least Squares (LS) and Recurrent Neural Networks (RNN)), and different training and validation sequences (intra subject "10 trials for all subjects and 1 trial for validation" and inter-subject "8 subjects for training and 1 for validation"). Furthermore, the authors have compared using the amputated hip data and both hips' data to determine the best set of input for knee and ankle angle prediction during LG and stair ascend (SA). Interestingly, the authors found that using both hips' data is redundant for trajectory prediction. The RNN had a slightly better performance compared to LS. Even though the lowest RMSE was obtained using intra-subject maps, the gait prediction with RNN using inter-subject maps has the best prediction on average across subjects with 6.15°, and 5.74° RMSE for the knee and ankle, respectively.

In [112–115], the controller consists of two recurrent long short-term memory neural networks (LSTM RNN) to estimate the ankle angle of the amputated leg-based on the kinematics of the intact and residual limbs. In [113], the authors added additional sensors and increased the number of neurons in the hidden layers and included the history of the ankle joints. The authors improved their system by adding egocentric vision using an eye tracker to get a better estimation. In [115], the control system consists of two LSTM networks but not as in their former structure. In this system, one neural network work received the 57 kinematic features of 19 intact joints as inputs and the other received the egocentric data from the eye tracker. Then, a fusion layer will merge the outputs of the neural networks to estimate the desired trajectories of both right knee and ankle angles in the sagittal plane. The reference control signals that the controllers generate were LG, and LG with obstacles [112–114]. The controller was able to generate the walking trajectory in LG with an average of 4.23° RMSE and <6.3° average RMSE for LG with obstacles. The overall RMSE of the developed method in [115] over four different walking terrains for the knee and ankle were 0.129 ± 0.057 rad and 0.119 ± 0.052 rad, respectively. The overall personal correlation coefficients (PCC) for the knee and ankle for the same walking terrains were 0.799 ± 0.205 and 0.637 ± 0.264, respectively. The egocentric vision improved the performance of both knee and ankle estimation compared to the previous methods by 7.9 and 7% of the RMSE, respectively.

In [116], A NARX RNN was developed to estimate the ankle angle using EMG signals from the residual limb muscles, LG, SA/stair descend (SD) and RA/ramp descend (RD) were the terrains on which the controller was trained and tested. The average RMSE for all subjects at variant terrains was 4.204° and the average correlation was 0.906. Moreover, the average cross-correlation coefficient for the hip, knee, and ankle for all walking speeds were 0.92 ± 0.06, 0.92 ± 0.07, and 0.86 ± 0.05, respectively.

In [117], authors have developed a wavelet neural network (WNN) for pattern generator. The WNN predict the joint angle based on the collected EMG signals. To generate the desired trajectory, the pre-processed EMG signals along with the joint angle were used as input signals of the WNN. To train the WNN, the authors used a first-order gradient decent method. The authors in this study represent a promising control concept for rehabilitation robots. However, their publication lacks of clear assessment of the results. The average error of the estimated trajectory by WNN of the subjects was not reported in the study. They only reported the tracking error of the MRAC method in the range of ± 0.01 m compared to the range of PID tracking error is ± 0.023 m.

A time delay feed-forward multilayer neural network was trained to estimate ankle angle and ankle moment using EMG signals as input data [118]. In this study, the authors have used the open database collected [119]. The best performance was obtained when the combination of the medial gastrocnemius (MG), biceps femoris (BF) and gluteus maximus (GM) EMG signals were used as inputs, which yielded in 1.53° and 0.04 Nm/kg RMSE, and over 98% correlation in position and moment estimations.

Authors in [120] developed a foot trajectory generator based on NARX RNN. The foot orientation of the missing limb is estimated based on the tibia angular velocity of the same leg, with the angular velocity measured by an IMU. The model was trained using the data of one able-bodied subject with one walking speed and tested on seven subjects while each walked at three different walking speeds (their normal walking speed and ±2% of their normal speed). The neural network was able to generate the desired foot orientation for all subjects at variant walking speeds with an average 2.1° ± 1.7° RMSE, and high correlation with the real foot trajectories (i.e., up to 98.8%). The algorithm was developed to work alongside an adaptive fuzzy system based on Takagi–Sugeno–Kang inference at the mid-level [121]. The output membership functions were designed to recreate an impedance control law based on the state of the gait cycle. The adaptive feature of the control system allows for automatic tuning and adjustable knowledgebase when the system faces unforeseen input fuzzy state. However, the system is now fully adaptive as the knowledgebase can only integrate new rules without eliminating or adaptively refining the knowledgebase. Moreover, the control system comes with high computational cost and the cost can increase with the introduction of adaptive new rules without eliminating the passive rules.

In [122], authors have integrated their reinforcement learning-based adaptive impedance controller [55, 60, 62] with a Bayesian human-in-the-loop (HIL) optimization in a hierarchical control scheme to improve the symmetry of transfemoral amputees' gait cycle. The HIL will provide the optimal features of the knee joint kinematics associated with desired human-prosthesis gait performance based on the symmetry index and the magnitude of four local extremum points of the previous knee joint gait. In other words, the optimization algorithm will converge the output to the points leading to the most symmetric gait cycle rather than generating a knee trajectory similar to an able-bodied trajectory. The proposed method is a developed version of the authors' actor-critic-based RL tracking control which makes the prosthetic knee mimic the intact knee kinematics [123]. After generating the optimal features, the RL learn an optimal policy to modify the FSM impedance parameters to generate a matching trajectory. To evaluate the proposed algorithm two able-bodied subjects and one transfemoral amputee have participated in an experiment. The results indicate that the developed method was capable of increasing the gait symmetry by an average of 50.23 ± 0.90% when the HIL generates the optimal knee features compared to the passive mode and by an average of 39.21 ± 10.24% compared to the worst knee features. Moreover, the algorithm leads to a greatest impulse, as the subjects' impulses reached 37.3 ± 6.3N·s, 26.4 ± 3.6N·s and 36.3 ± 4.4 N·s for the two able-bodied subjects, and the transfemoral amputee subject, respectively.

The proposed system in [81] relied on 250 ms history of the vertical load, thigh angle, knee angle, and ankle angle to generate the continuous gait phase. The deep neural network (DNN) consists of a convolutional layer (50,32), LSTM (20), and two fully connected layers (30 and 10). In [82], authors have proposed a generative adversarial network (GAN) to generate synthetic gait patterns. The generated pattern was used to train the DNN model control developed [80], which generates the impedance parameters throughout the gait cycle. The same input data were used to generate the required pattern. The average R^2 for the able-bodied data between the predicted and the measured trajectories was 0.97, and the RMSE for the thigh, knee, and ankle angles were 6.2∘, 4.3∘, and 3.7∘ for LG, and 4.3∘, 8.3∘, and 1.9∘ for SA, respectively. The average R^2 between the predicted phase and the actual one for all amputee users was 0.72. The R^2 between the goniometer data and corresponding synthetic gait patterns values were 0.99, 0.99, 0.99, and 0.91, respectively. The RMSE values were 0.16°, 0.66°, 1.15°, and 2.38°, respectively. The GAN was able to generate standing-to-sitting and sitting-to-standing without being trained on such patterns.

The controller in [79] relied on the non-normalized EMG signals from the residual limb to estimate the ankle position and moment. Two LSTM RNNs were trained to map the relation between EMG signals and the ankle position and moment. The results showed that the most accurate prediction was obtained from integrated EMG and waveform length features, and the best muscle combination in terms of prediction accuracy was medial gastrocnemius, rectus femoris, and vastus medialis ($r_{position} = 0.9099$, $r_{moment} = 0.9707$, $RMSE_{position} = 4.5923° \pm 1.4815$, $RMSE_{moment} = 0.1072$ N · m/kg ± 0.0660). But in terms of the economic and practical variations, peroneus longus ($r_{position} = 0.9001$, $r_{moment} = 0.9703$) and gluteus maximus and vastus medialis ($r_{position} = 0.9010$, $r_{moment} = 0.9718$, $RMSE_{position} = 4.8927° \pm 1.6140$, $RMSE_{moment}$ 0.1193 N · m/kg ± 0.0501) found to be the best.

5 The Future of Lower Limb Prostheses: Internet of Things Integration and Personalized Care

The Internet of Things (IoT) is a transformative technological paradigm that has ushered in an era of unprecedented connectivity and data exchange. It emerges as a powerful domain with key enablers such as processors and simple-to-setup wireless networks [124]. IoT is a system made up of real-world objects with embedded sensors, software, electronics, and network connectivity. The connectivity empowers these objects to gather and exchange data while also can be monitored and controlled remotely through a network infrastructure that is already in place [125].

The impact of IoT extends beyond enhanced independence, it also enriches human interaction with the external environment. With the aid of futuristic protocols and algorithms, IoT has emerged as a major component of global connection [126]. It is being utilized in a variety of settings, from healthcare to smart homes and industrial automation [127].The widespread popularity of IoT is driven by its capacity to provide higher accuracy, lower costs, and enhanced predictive capabilities. Furthermore, this rapid IoT revolution is further fueled by increased software knowledge and applications, alongside advancements in computer and mobile technologies. The easy accessibility of wireless technology and the growing digital economy have also propelled the IoT into the forefront of technological innovation [128].

IoT technology has wrought a profound transformation in the medical sector. While it cannot halt the ageing process or eradicate chronic diseases, it contributes significantly to enhancing accessibility and offering innovative solutions to long-standing challenges. IoT enhances well-being and healthcare through various means such as streamlined communication, remote monitoring, task automation and facilitating access to a wide array of healthcare services and information [129]. This technological advancement has ushered in new opportunities for personalized healthcare services, endowing them with advantages like enhanced accessibility, greater availability, tailored content, and cost-efficient delivery [130].

The IoT devices ensure that patients receive timely care by continually collecting and transmitting data. For example, smart healthcare with sensors and actuators autonomously collects and transmits the data in real-time to servers with little or no human involvement [124]. It makes physicians' and patients' lives more convenient as it enables healthcare professionals to take appropriate actions according to the data collected [125, 131].

Furthermore, IoT technologies have been widely used in the design and implementation of various medical equipment employed for testing, monitoring, diagnosis, and treatment [132]. For instance, authors in [133] have proposed the concept of wireless body area networks (WBAN) to cater to the demands of remote health care. Through wearable sensors on the body or implanted sensors within the body, the physiological parameters including heart rate, pulse rate, blood pressure, body temperature, electrocardiogram (ECG), and electroencephalogram (EEG) are collected, monitored, and transmitted to medical centers. This allows doctors to remotely diagnose potential diseases and propose treatment plans. This not only eases the burden on patients to attend medical centers for check-ups but also conserves valuable medical resources.

Additionally, the interconnected IoT healthcare devices hold immense potential to significantly improve patient care. This transformation has shifted the healthcare paradigm from a hospital-centric system to a patient-centric one, where patients assume a more active role in managing their own health [134, 135]. Patients are able to continually track the progression of their health status and cooperate with healthcare providers to personalize care and treatment based on their completed health targets [136].

The field of prosthetics and orthotics are no exception to the transformative potential of IoT technology. By connecting prosthetic devices to the internet and integrating them into the IoT, their functionality can be enhanced. IoT technology enables the collection of comprehensive details to process the data, paving the way for multifunctional prosthetic devices. For instance, a mind-controlled robotic arm with a smart sensor network has been developed, including temperature sensors, pressure sensors, accelerometers, gyroscopes, potentiometers, and proximity sensors. EEG signals were used to control the robotic arm, while the smart sensors facilitated interactions with the surrounding environment [137]. Other than that, the Ubi-Sleeve that was designed by [138], allowed individuals with lower-limb prostheses and other stakeholders to regularly capture signals like humidity, temperature, and prosthesis slippage behavior from within the socket. Besides, the implementation of IoT in prostheses also includes the development of exoskeleton robots for lower limbs that incorporate efficient mass adjustment mechanisms to ensure stability when moving [139]. These innovative approaches hold the potential to revolutionize prosthetic care and research.

The current state of prosthetic technology does not fully address user needs. Ineffective channels of communication between clinicians, users, and manufacturers impede the acquisition of qualitative and quantitative information required for improving prosthetic innovations [123]. According to the authors in [123], the current clinical practice makes it challenging for clinicians to remotely monitor prosthesis usage and make necessary adjustments. The absence of regular maintenance can lead to reduced comfort and functionality, ultimately contributing to a higher rate of prosthesis abandonment. User feedback is typically collected through email or telephone calls, which have limitations in their effectiveness. Moreover, users struggle to provide contextual information about their prosthesis usage accurately.

To deal with this situation, the IoT presents an opportunity for real-time visualization, remote data collection, and reprogramming of prostheses via Bluetooth and cloud portals with the development of smart foot prostheses. Research has demonstrated that active prosthetic devices hold the promise of providing greater comfort and reduced energy expenditure for amputees when compared to their passive counterparts [140]. The synergy between prostheses and the IoT is believed to bring advancing adaptability and resemble the plasticity of intact lower limbs. This integration offers new possibilities for improving the quality of life and functionality of individuals with limb loss [141].

Smart prosthetic devices can also play a crucial role in ensuring effective rehabilitation practices and making informed clinical decisions. Monitoring the device is crucial for elderly prosthetic users as they are often at a higher risk of falling [142]. By detecting instability or irregular movement patterns in the prosthesis, the potential fall risk can be identified. The clinicians can be promptly alerted and take preventive actions. Besides, monitoring the foot prosthesis can determine if the patient is an active user. The information is important in rehabilitation settings where maintaining physical activity is essential.

5.1 Remote Monitoring with Smart Prostheses

The traditional monitoring approach for the use of prosthetic devices is subjective. It focused on interviews, surveys, or diaries [143, 144]. However, subjective methods of data gathering may not adequately capture a patient's usage of a prosthesis since they are frequently constrained by elements like generalization, comprehension, perception, honesty, and memory [145]. Although prosthesis users may receive training in a clinic, it is uncertain what happens to them once they get home. Moreover, as a result of the recent emphasis on evidence-based approaches, standardized measures are being used more often to assess results, monitor progress and facilitate communication between patients and prosthetists [146].

The quality of care given to patients with amputation could be greatly enhanced by the use of sensing and monitoring technologies. Using these technologies in the rehabilitation process opens up possibilities for the multimodal exchange of relevant and useful health information between patients, prosthetic devices, and medical professionals [147, 148]. In a pilot study conducted [149], the perceptions of prosthesis use and the activity data were obtained from the prosthetists and monitoring device respectively. The result showed that prosthetists may be able to gain a better understanding of their patients' prosthesis use and activity with the use of monitors. The monitoring system's data may guide clinical judgements and enhance evidence-based procedures.

Remote monitoring systems in prosthetic devices allow for continuous evaluation of prosthetic function and residual limb health. Hafner and Sanders [148] mentioned that even with a prosthesis, people who have lost their lower limbs will face a lifetime of functional restrictions and ongoing health risks. These include issues with higher metabolic energy use, decreased activity level, slow walking speed, reduced balance, asymmetries in gait, increased risk of stumbling and falling, trouble in navigating stairs and uneven terrain, pain, skin breakdown and joint degeneration.

By designing a wearable sensor system that can gather data from sensors, store and transmit the data to a central health database for analysis, long-term monitoring can be achieved and bring up limitless possibilities [150] (Fig. 3). Benefits of remote monitoring include lowering clinical visitation rates and hospitalization rates, lowering medical expenses and improving life quality [151].

Besides, authors in [152] developed a non-invasive monitoring technology to monitor prosthetic users' activity and doffing by using accelerometers and proximity sensors. Proximity sensors were placed on the medial and lateral socket rims, while commercial accelerometers were placed on the thigh and lower leg. Data collection was transformed by a unique data-processing algorithm into a time-stamped list of postures and activities: sitting, standing, doffed, and walking (including all forms of movement).

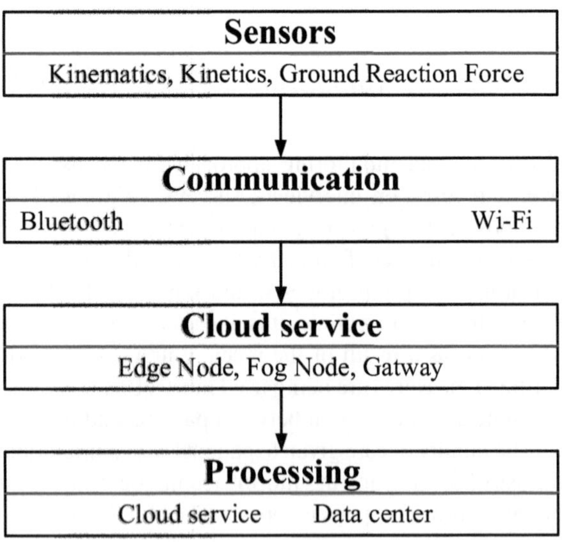

Fig. 3 Smart prostheses with IoT capabilities and layers

Additionally, remote monitoring using sensors enables remote evaluation and early detection of emergency cases, contributing to psychological and physical rehabilitation. For example, by utilizing the sensor data, we are able to identify the patient's motions and establish thresholds that may signal potentially harmful actions like slipping or falling [153]. Moreover, evaluation of body movements helps to monitor mobility. Proper sensors facilitate the rehabilitation process and maximize patients' recovery by improving exercise techniques through movement evaluation [154].

5.1.1 Data Acquisition

In the field of prosthetics and orthotics, data acquisition can be achieved by connecting various sensors. Sensors play a crucial role in enhancing the functionality of prostheses. Authors in [155] proposed a wearable sensor platform to remotely evaluate prosthetic lower limbs. The sensors were placed on the prosthesis. Sensors used include a force sensor to measure pressure distribution during walking, EMG sensors and temperature sensors MLX90614 to monitor muscle activity and temperature, variable resistance and gyroscope to evaluate the amount of opening and closing angle of the prosthesis and inclination, accelerometers for acceleration.

Biosensors are widely used to assess kinetics and spatial and temporal features to facilitate active tracking of amputee prosthetics systems [147, 155]. The data collected is crucial for monitoring the status of the prosthetic limb, ensuring the proper functioning of the prosthesis and detecting any abnormalities.

The paper presented by [156], assessed the accuracy and reliability of an IMU system in comparison to an optical motion capture system among individuals using transtibial prostheses during walking. The results show good to excellent agreement between the two systems for all sagittal kinematic parameters.

Another article by [157], proposes a method for real-time limb tracking in single-depth images using IMU data. The method combines a deep neural network with a Kalman filter to estimate the position and orientation of the prosthetic leg in a world reference system. The method is tested on a dataset of lower limb prosthesis users performing various activities.

Moreover, authors in [57] present a method for intelligent lower-limb prosthesis control by using IMU and EMG signals. The method uses locally weighted projection regression (LWPR) to learn the relationship between the sensor signals and the desired prosthesis state. The method is evaluated on a simulated prosthesis model and shows promising results in terms of accuracy and adaptability.

Sensor data accuracy and reliability are vital for ensuring safety and stability, enhancing performance in prosthetic control and comfortableness as well as providing evaluation and feedback that allow clinicians, researchers, and prosthesis manufacturers to improve the design, provision, and service of prosthetic devices [147]. Sensor data can help detect and prevent falls, slips, or collisions by adjusting the prosthesis accordingly. Accurate and reliable sensor data can also reduce the risk of injury or damage to the prosthesis or the user, and improve the user's gait quality, energy efficiency, and satisfaction [158, 159].

5.1.2 Data Transmission

Data transmission in smart prostheses is the process of transmitting and receiving information between the prosthetic device and the user's body or external devices. Data transmission enables the user to control the prosthesis using signals from the brain, muscles, nerves or other sources, provides the user with sensory feedback from the prosthesis such as touch, pressure, temperature, or pain as well as monitors the usage and performance of prosthesis such as activity level, socket fit, temperature, humidity, or battery status [147, 160, 161].

Data transmission protocol sets the rules or standards that govern how data is exchanged between devices or systems. In the context of smart prosthesis, data transmission protocol is important for enabling communication between the prosthesis and the user, the prosthesis and the external environment, and the prosthesis and the clinician or technician [161]. Data transmission is shifting from wired to wireless with innovations like Bluetooth and Wi-Fi [162].

Bluetooth is a wireless technology that uses short-range radio waves to connect devices. Bluetooth is widely used for smart prostheses because it is low-cost, low-power, and compatible with many devices such as smartphones, tablets, and computers. Bluetooth can support data rates up to 24 Mbps and data reliability up to 99.9%. However, Bluetooth has limited range (up to 100 m) and security (susceptible to eavesdropping and hacking) [147, 163].

Wi-Fi is a wireless technology that uses radio waves to connect devices to a local area network (LAN) or the internet. Wi-Fi is also used for smart prostheses because it can provide high data rate (up to 600 Mbps), high data reliability (up to 99.999%), and high data security (using encryption and authentication). However, Wi-Fi has higher power consumption and requires access to a wireless router or hotspot, which may not be available in some environments [164].

ZigBee is a wireless technology that uses low-power radio waves to create a mesh network of devices. ZigBee is suitable for smart prostheses because it can support low data rates (up to 250 kbps), low power consumption (up to 10 years of battery life), and high data reliability (up to 99.999%). ZigBee can also support long-range (up to 10 km) and high data security (using encryption and authentication). However, ZigBee has limited compatibility and interoperability with other devices and protocols [165].

Bluetooth Low Energy (BLE), also known as Bluetooth Smart, is a wireless communication technology designed for short-range communication with low power consumption. It is part of the Bluetooth 4.0 core specification and is intended to provide energy-efficient connectivity for devices that need to operate for extended periods on small batteries, such as wearable devices, smart home products, and medical devices.

6 Conclusion

Advances in wearable technologies along with the AI evolution are transforming the field of lower limb prostheses, bringing positive impacts in prosthetic design, functionality, customization, and user experience. Gait assessment using the AI-based approaches provides detailed insights into prosthetic gait and body movements, easing adjustments of the lower limb prosthesis and personalization of rehabilitation strategies. AI-enabled powered prosthesis produces more efficient gait and intuitive limb movements through an advanced control system, which greatly improves mobility and the quality of life. Future lower limb prosthesis assures integration with other smart devices and systems. AI will continue to enhance prosthetic functionality through user-device communication, allowing remote monitoring, personalized adjustments and finally improved healthcare management.

References

1. Duysens, J.: Human gait as a step in evolution. Brain, 2589–2590 (2002)
2. Perry, J.: Gait Analysis: Normal and Pathological Function. Slack Incorporated, New Jersey (2010)
3. McDonald, C.L., Westcott-McCoy, S., Weaver, M.R., Haagsma, J., Kartin, D.: Global prevalence of traumatic non-fatal limb amputation. Prosthet. Orthot. Int. 105–114 (2021)
4. Rogers, J.: McCance and Huether's Pathophysiology: The Biologic Basis for Disease in Adults and Children, 9th edn. Mosby, Missouri (2022)
5. Sun, H., Saeedi, P., Karuranga, S., Pinkepank, M., Ogurtsova, K., Duncan, B.B., Stein, C., Basit, A., Chan, J.C.N., Mbanya, J.C., Pavkov, M.E., Ramachandaran, A., Wild, S.H., James, S., Herman, W.H., Zhang, P., Bommer, C., Kuo, S., Boyko, E.J., Magliano, D.J.: IDF diabetes atlas: global, regional and country-level diabetes prevalence estimates for 2021 and projections for 2045. Diab. Res. Clin. Pract. 109119 (2022)
6. Ezzatvar, Y., García-Hermoso, A.: Global estimates of diabetes-related amputations incidence in 2010–2020: a systematic review and meta-analysis. Diab. Res. Clin. Pract. 110194 (2023)
7. Breakey, J.: Gait of unilateral below-knee amputees. Orthot. Prosthet. 17–24 (1976). https://doi.org/10.1016/j.diabres.2022.110194
8. Culham, E.G., Peat, M., Newell, E.: Below-knee amputation: a comparison of the effect of the SACH foot and single axis foot on electromyographic patterns during locomotion. Prosthet. Orthot. Int. 15–22 (1986)
9. Doane, N.E., Holt, L.E.: A comparison of the SACH and single axis foot in the gait of unilateral below-knee amputees. Prosthet. Orthot. Int. 33–36 (1983)
10. Engsberg, J.R., Lee, A.G., Patterson, J.L., Harder, J.A.: External loading comparisons between able-bodied and below-knee-amputee children during walking. Arch. Phys. Med. Rehabil. 657–661 (1991)
11. Bateni, H., Olney, S.J.: Kinematic and kinetic variations of below-knee amputee gait. J. Prosthet. Orthot. 2–10 (2002)
12. Saleh, M., Murdoch, G.: In defence of gait analysis. Observation and measurement in gait assessment. J. Bone Joint Surg., Br. **67**(2), 237–241 (1985)
13. Gard, S.A.: Use of quantitative gait analysis for the evaluation of prosthetic walking performance. J. Prosthet. Orthot. 93–104 (2006)
14. Esquenazi, A.: Analysis of prosthetic gait. In: Esquenazi, A. (ed.) Physical Medicine and Rehabilitation State of the art Reviews, pp. 201–220. Henley & Belfus Inc., Philadelphia (1994)

15. Esquenazi, A., Wikoff, E., Lucas, M.: Amputation rehabilitation. In: Grabois, M. (ed.) Physical Medicine and Rehabilitation—The Complete Approach, pp. 1744–1760. Blackwell Science, Oxford (2000)
16. Eastlack, M., Arvidson, J., Snyder-Mackler, L., Danoff, J.V., McGarvey, C.L.: Interrater reliability of videotaped observational gait analysis. Phys. Theraphy. **71**(6), 465–472 (1991)
17. Mohan, D.M., Khandoker, A.H., Wasti, S.A., Ismail Ibrahim Ismail Alali, S., Jelinek, H.F., Khalaf, K.: Assessment methods of post-stroke gait: a scoping review of technology-driven approaches to gait characterization and analysis. Front. Neurol. (2021)
18. Esquenazi, A.: Gait analysis in lower-limb amputation and prosthetic rehabilitation. Phys. Med. Rehabil. Clin. North Am. 153–167 (2014)
19. Richards, J., Levine, D., Whittle, M.W.: Whittle's Gait Analysis, 6th edn. Elsevier, Amsterdam (2023)
20. Favre, J., Jolles, B.M., Aissaoui, R., Aminian, K.: Ambulatory measurement of 3D knee joint angle. J. Biomech. 1029–1035 (2008)
21. Takeda, R., Tadano, S., Natorigawa, A., Todoh, M., Yoshinari, S.: Gait posture estimation using wearable acceleration and gyro sensor. J. Biomech. 2486–2494 (2009)
22. Liu, K., Liu, T., Shibata, K., Inoue, Y., Zheng, R.: Novel approach to ambulatory assessment of human segmental orientation on a wearable sensor system. J. Biomech. 2747–2752 (2009)
23. Tucker, M.R., Olivier, J., Pagel, A., Bleuler, H., Bouri, M., Lambercy, O., Del Millán, J.R., Riener, R., Vallery, H., Gassert, R.: Control strategies for active lower extremity prosthetics and orthotics: a review. J. NeuroEng. Rehabil. 1–30 (2015)
24. Tham, L.K., Abu Osman, N.A., Al-Kouzbary, M., Aminian, K.: Biomechanical ambulatory assessment of 3D knee angle using novel inertial sensor-based technique. IEEE Access 36559–36570 (2021)
25. Vu H.T.T., Dong, D., Cao, H.-L., Verstraten, T., Lefeber, D., Vanderborght, B., Geeroms, J.: A review of gait phase detection algorithms for lower limb prostheses. Sensors (Basel) 3972 (2020). https://doi.org/10.3390/s20143972
26. Kouzbary, H.A., Kouzbary, M.A., Tham, L.K., Liu, J., Shasmin, H.N., Abu Osman, N.A.: Generating an adaptive and robust walking pattern for a prosthetic ankle-foot by utilizing a nonlinear autoregressive network with exogenous inputs. IEEE Trans. Neural Netw. Learn. Syst. 6297–6305 (2022)
27. Sánchez Manchola, M.D., Bernal, M.J.P., Munera, M., Cifuentes, C.A.: Gait phase detection for lower-limb exoskeletons using foot motion data from a single inertial measurement unit in Hemiparetic individuals. Sensors 2988 (2019). https://doi.org/10.3390/s19132988
28. Jung, J.-Y., Heo, W., Yang, H., Park, H.: A neural network-based gait phase classification method using sensors equipped on lower limb exoskeleton robots. Sensors (Basel) 27738–27759 (2015)
29. Nazmi, N., Abdul Rahman, M.A., Yamamoto, S.-I., Ahmad, S.A.: Walking gait event detection based on electromyography signals using artificial neural network. Biomed. Signal Process. Control 334–343 (2019). https://doi.org/10.1016/j.bspc.2018.08.030
30. Ma, Y., Wu, X., Wang, C., Yi, Z., Liang, G.: Gait phase classification and assist torque prediction for a lower limb exoskeleton system using Kernel Recursive Least-Squares method. Sensors 5449 (2019). https://doi.org/10.3390/s19245449
31. Zhen, T., Yan, L., Yuan, P.: Walking gait phase detection based on acceleration signals using LSTM-DNN algorithm. Algorithms 253 (2019). https://doi.org/10.3390/a12120253
32. Tham, L.K., Al Kouzbary, M., Al Kouzbary, H., Liu, J., Abu Osman, N.A.: Estimation of body segmental orientation for prosthetic gait using a nonlinear autoregressive neural network with exogenous inputs. Phys. Eng. Sci. Med. 1723–1739 (2023)
33. Ng, G., Andrysek, J.: Classifying changes in amputee gait following physiotherapy using machine learning and continuous inertial sensor signals (2023). https://doi.org/10.3390/s23031412
34. Rathore, R., Singh, A.K., Chaudhary, H., Kandan, K.: Gait abnormality detection in unilateral trans-tibial amputee in real-time gait using wearable setup. IEEE Sens. J. 12567–12573 (2023). https://doi.org/10.1109/JSEN.2023.3263399

35. Juneau, P., Baddour, N., Burger, H., Bavec, A., Lemaire, E.D.: Amputee fall risk classification using machine learning and smartphone sensor data from 2-minute and 6-minute walk tests. Sensors 1749 (2022). https://doi.org/10.3390/s22051749
36. Wang, X., Meng, Q., Bai, S., Meng, Q., Yu, H.: Hybrid active–passive prosthetic knee: a gait kinematics and muscle activity comparison with mechanical and microprocessor-controlled passive prostheses. J. Bionic Eng. 119–135 (2023)
37. Goldfarb, M., Lawson, B.E., Shultz, A.H.: Realizing the promise of robotic leg prostheses. Sci. Transl. Me. **5**(210), 210 ps15-ps15 (2013)
38. Waters, R., Perry, J., Antonelli, D., Hislop, H.: Energy cost of walking of amputees: the influence of level of amputation. J. Bone Joint Surg. Am. **58**(1), 42–46 (1976)
39. Ephraim, P.L., Wegener, S.T., MacKenzie, E.J., Dillingham, T.R., Pezzin, L.E.: Phantom pain, residual limb pain, and back pain in amputees: results of a national survey. Arch. Phys. Med. Rehabil. **86**(10), 1910–1919 (2005)
40. Burke, M., Roman, V., Wright, V.: Bone and joint changes in lower limb amputees. Ann. Rheum. Dis. **37**(3), 252–254 (1987)
41. Lemaire, E.D., Fisher, F.R.: Osteoarthritis and elderly amputee gait. Arch. Phys. Med. Rehabil. **5**(10), 1094–1099 (1994)
42. Runciman, P., Cockcroft, J., Derman, W.: A novel pivot ankle/foot prosthesis reduces sound side loading and risk for osteoarthritis: a pragmatic randomized controlled trial. Prosthet. Orthot. Int. **46**(3), 258 (2022)
43. Horgan, O., MacLachlan, M.: Psychosocial adjustment to lower-limb amputation: a review. Disabil. Rehabil. **26**(14–15), 837–850 (2004)
44. Behel, J.M., Rybarczyk, B., Elliott, T.R., Nicholas, J.J., Nyenhuis, D.: The role of perceived vulnerability in adjustment to lower extremity amputation: a preliminary investigation. Rehabil. Psychol. **47**(1), 92 (2002)
45. Jones, L., Hall, M., Schuld, W.: Ability or disability? A study of the functional outcome of 65 consecutive lower limb amputees treated at the Royal South Sydney Hospital in 1988–1989. Disabil. Rehabil. **15**(4), 184–188 (1993)
46. Jimenez-Fabian, R., Verlinden, O.: Review of control algorithms for robotic ankle systems in lower-limb orthoses, prostheses, and exoskeletons. Med. Eng. Phys. **34**(4), 397–408 (2012)
47. Jayaraman, C., Mummidisetty, C.K., Albert, M.V., Lipschutz, R., Hoppe-Ludwig, S., Mathur, G., Jayaraman, A.: Using a microprocessor knee (C-Leg) with appropriate foot transitioned individuals with dysvascular transfemoral amputations to higher performance levels: a longitudinal randomized clinical trial. J. Neuroeng. Rehabil. **18**(1), 1–13 (2021)
48. Windrich, M., Grimmer, M., Christ, O., Rinderknecht, S., Beckerle, P.: Active lower limb prosthetics: a systematic review of design issues and solutions. Biomed. Eng. Online **15**(3), 5–19 (2016)
49. Grimes, D., Flowers, W.C., Donath, M.: Feasibility of an active control scheme for above knee prostheses (1977)
50. Prasanna, C., Realmuto, J., Anderson, A., Rombokas, E., Klute, G.: A data-driven and personalized stance symmetry controller for robotic ankle-foot prostheses: a preliminary investigation. IEEE Trans. Neural Syst. Rehabil. Eng. **31**, 4051–4062 (2023). https://doi.org/10.1109/TNSRE.2023.3322124
51. Realmuto, J., Klute, G., Devasia, S. (eds.): Preliminary investigation of symmetry learning control for powered ankle-foot prostheses. In: 2019 Wearable Robotics Association Conference (WearRAcon). IEEE (2019)
52. Alili, A., Nalam, V., Li, M., Liu, M., Si, J., Huang, H.H. (eds.): User controlled interface for tuning robotic knee prosthesis. In: 2021 IEEE/RSJ International Conference on Intelligent Robots and Systems (IROS). IEEE (2021)
53. Gao, X., Si, J., Wen, Y., Li, M., Huang, H.: Reinforcement learning control of robotic knee with human-in-the-loop by flexible policy iteration. IEEE Trans. Neural Netw. Learn. Syst. (2021)
54. Huang, H., Crouch, D.L., Liu, M., Sawicki, G.S., Wang, D.: A cyber expert system for auto-tuning powered prosthesis impedance control parameters. Ann. Biomed. Eng. **44**, 1613–1624 (2016)

55. Li, M., Gao, X., Wen, Y., Si, J., Huang, H.H. (eds.): Offline policy iteration based reinforcement learning controller for online robotic knee prosthesis parameter tuning. In: 2019 International Conference on Robotics and Automation (ICRA). IEEE (2019)
56. Liu, M., Zhang, F., Datseris, P., Huang, H.H.: Improving finite state impedance control of active-transfemoral prosthesis using dempster-shafer based state transition rules. J. Intell. Rob. Syst. **76**(3), 461–474 (2014)
57. Liu, Y., An, H., Ma, H., Wei, Q.: An LWPR-based method for intelligent lower-limb prosthesis control by learning the dynamic model in real time. Machines [Internet] **11**(2) (2023). https://doi.org/10.3390/machines11020186
58. Wang, D., Liu, M., Zhang, F., Huang, H. (eds.): Design of an expert system to automatically calibrate impedance control for powered knee prostheses. In: 2013 IEEE 13th International Conference on Rehabilitation Robotics (ICORR). IEEE (2013)
59. Wen, Y., Brandt, A., Liu, M., Huang, H., Si, J. (eds.): Comparing parallel and sequential control parameter tuning for a powered knee prosthesis. In: 2017 IEEE International Conference on Systems, Man, and Cybernetics (SMC). IEEE (2017)
60. Wen, Y., Li, M., Si, J., Huang, H.: Wearer-prosthesis interaction for symmetrical gait: a study enabled by reinforcement learning prosthesis control. IEEE Trans. Neural Syst. Rehabil. Eng. **28**(4), 904–913 (2020)
61. Wen, Y., Liu, M., Si, J., Huang, H.H. (eds.): Adaptive control of powered transfemoral prostheses based on adaptive dynamic programming. In: 2016 38th Annual International Conference of the IEEE Engineering in Medicine and Biology Society (EMBC). IEEE (2016)
62. Wen, Y., Si, J., Brandt, A., Gao, X., Huang, H.H.: Online reinforcement learning control for the personalization of a robotic knee prosthesis. IEEE Trans Cybern. **50**(6), 2346–2356 (2019)
63. Wen, Y., Si, J., Gao, X., Huang, S., Huang, H.H.: A new powered lower limb prosthesis control framework based on adaptive dynamic programming. IEEE Trans. Neural Netw. Learn. Syst. **28**(9), 2215–2220 (2016)
64. Wu, R., Yao, Z., Si, J., Huang, H.H.: Robotic knee tracking control to mimic the intact human knee profile based on actor-critic reinforcement learning. IEEE/CAA J. Automatica Sinica. **9**(1), 19–30 (2021)
65. Bhakta, K., Camargo, J., Donovan, L., Herrin, K., Young, A.: Machine learning model comparisons of user independent & dependent intent recognition systems for powered prostheses. IEEE Robot. Autom. Lett. **5**(4), 5393–5400 (2020)
66. Contreras-Cruz, M.A., Novo-Torres, L., Villarreal, D.J., Ramirez-Paredes, J.-P.: Convolutional neural network and sensor fusion for obstacle classification in the context of powered prosthetic leg applications. Comput. Electr. Eng. **108**, 108656 (2023)
67. Hu, B., Rouse, E., Hargrove, L.: Fusion of bilateral lower-limb neuromechanical signals improves prediction of locomotor activities. Front. Robot. AI **5**, 78 (2018)
68. Liu, L., Yang, P., Liu, Z., Song, Y.: Prosthetic leg locomotion-mode identification based on high-order zero-crossing analysis surface electromyography. J. Shanghai Jiaotong Univ. (Sci.) **26**(1), 84–92 (2021)
69. Murray, R., Mendez, J., Gabert, L., Fey, N.P., Liu, H., Lenzi, T.: Ambulation mode classification of individuals with transfemoral amputation through A-mode sonomyography and convolutional neural networks. Sensors **22**(23), 9350 (2022)
70. Papapicco, V., Chen, B., Munih, M., Davalli, A., Sacchetti, R., Gruppioni, E., Crea, S., Vitiello, N.: A classification approach based on directed acyclic graph to predict locomotion activities with one inertial sensor on the thigh. IEEE Trans. Med. Robot. Bionics **3**(2), 436–445 (2021)
71. Song, Y., Zhu, Y., Zheng, E., Tao, F., Wang, Q. (eds.): Optimizing support vector machine with genetic algorithm for capacitive sensing-based locomotion mode recognition. In: Intelligent Autonomous Systems 13: Proceedings of the 13th International Conference IAS-13. Springer (2016)
72. Zhang, Y., Wang, X., Xiu, H., Ren, L., Han, Y., Ma, Y., Chen, W., Wei, G., Ren, L.: An optimization system for intent recognition based on an improved KNN algorithm with minimal feature set for powered knee prosthesis. J. Bionic Eng. 1–14 (2023)

73. Ma, T., Wang, Y., Chen, X., Chen, C., Hou, Z., Yu, H., Fu, C.: A piecewise monotonic smooth phase variable for speed-adaptation control of powered knee-ankle prostheses. IEEE Robot. Autom. Lett. **7**(3), 8526–8533 (2022)
74. Xia, Y., Ma, H., An, H., Huang, Y., Wei, Q. (eds.) The algebraic curve of virtual constraint relation in the control of lower limb prosthesis with different slope. In: 2020 Chinese Automation Congress (CAC). IEEE (2020)
75. Clites, T.R., Carty, M.J., Ullauri, J.B., Carney, M.E., Mooney, L.M., Duval, J.-F., Srinivasan, S.S., Herr, H.M.: Proprioception from a neurally controlled lower-extremity prosthesis. Sci. Transl. Med. **10**(443), 8373 (2018)
76. Hooda, N., Kumar, N.: Optimal channel-set and feature-set assessment for foot movement based EMG pattern recognition. Appl. Artif. Intell. **35**(15), 1685–1707 (2021)
77. Pi, M., Yuan, Y., Li, Z. (eds.): Control of the robotic leg prostheses based on a neuromuscular model. In: 2019 IEEE 4th International Conference on Advanced Robotics and Mechatronics (ICARM). IEEE (2019)
78. Wu, S.-K., Waycaster, G., Shen, X.: Electromyography-based control of active above-knee prostheses. Control Eng. Pract. **19**(8), 875–882 (2011)
79. Keleş, A.D., Türksoy, R.T., Yucesoy, C.A.: The use of nonnormalized surface EMG and feature inputs for LSTM-based powered ankle prosthesis control algorithm development. Front. Neurosci. **17** (2023)
80. Kim, M., Hargrove, L.J.: Deep-learning to map a benchmark dataset of non-amputee ambulation for controlling an open source bionic leg. IEEE Robot. Autom. Lett. **7**(4), 10597–10604 (2022)
81. Kim, M., Hargrove, L.J.: A gait phase prediction model trained on benchmark datasets for evaluating a controller for prosthetic legs. Front. Neurorobot. **16**, 1064313 (2023)
82. Kim, M., Hargrove, L.J.: Generating synthetic gait patterns based on benchmark datasets for controlling prosthetic legs. J. Neuroeng. Rehabil. **20**(1), 115 (2023)
83. Tomovic, R., McGhee, R.B.: A finite state approach to the synthesis of bioengineering control systems. IEEE Trans. Human Fact. Electron. **2**, 65–69 (1966)
84. Ju, M.-S., Yi, S.-H., Tsuei, Y.-G., Chou, Y.-L.: Fuzzy control of electrohydraulic above-knee prostheses. JSME Int. J. Ser. C, Dyn. Control Robot. Des. Manuf. **38**(1), 78–86 (1995)
85. Wiem, A., Rahma, B., Naoui, S.B.H.A. (eds.): An adaptive TS fuzzy-PID controller applied to an active knee prosthesis based on Human walk. In: 2018 International Conference on Advanced Systems and Electric Technologies (IC_ASET). IEEE (2018)
86. Tileylioğlu, E., Yilmaz, A. (eds.): Application of neural based estimation algorithm for gait phases of above knee prosthesis. In: 2015 37th Annual International Conference of the IEEE Engineering in Medicine and Biology Society (EMBC). IEEE (2015)
87. Li, M., Wen, Y., Gao, X., Si, J., Huang, H.: Toward expedited impedance tuning of a robotic prosthesis for personalized gait assistance by reinforcement learning control. IEEE Trans. Rob. **38**(1), 407–420 (2022). https://doi.org/10.1109/TRO.2021.3078317
88. Jannati, S., Yousefi-Koma, A., Ayati, M., Rezaeian, S. (eds.): Impedance control of a knee prosthesis with frictional torque estimation. In: 2019 7th International Conference on Robotics and Mechatronics (ICRoM). IEEE (2019)
89. Ha, K.H., Varol, H.A., Goldfarb, M.: Volitional control of a prosthetic knee using surface electromyography. IEEE Trans. Biomed. Eng. **58**(1), 144–151 (2010)
90. Ha, K.H., Varol, H.A., Goldfarb, M. (eds.): Myoelectric control of a powered knee prosthesis for volitional movement during non-weight-bearing activities. In: 2010 Annual International Conference of the IEEE Engineering in Medicine and Biology. IEEE (2010)
91. Tkach, D., Lipschutz, R.D., Finucane, S.B., Hargrove, L.J. (eds.): Myoelectric neural interface enables accurate control of a virtual multiple degree-of-freedom foot-ankle prosthesis. In: 2013 IEEE 13th International Conference on Rehabilitation Robotics (ICORR). IEEE (2013)
92. Garikayi, T., Van den Heever, D., Matope, S.: Analysis of surface electromyography signal features on osteomyoplastic transtibial amputees for pattern recognition control architectures. Biomed. Signal Process. Control **40**, 10–22 (2018)

93. Nandi, G.C., Ijspeert, A., Nandi, A. (eds.): Biologically inspired CPG based above knee active prosthesis. In: 2008 IEEE/RSJ International Conference on Intelligent Robots and Systems. IEEE (2008)
94. Nandi, G.C., Ijspeert, A.J., Chakraborty, P., Nandi, A.: Development of adaptive modular active leg (AMAL) using bipedal robotics technology. Robot. Auton. Syst. **57**(6–7), 603–616 (2009)
95. Delis, A.L., da Rocha, A.F., Dos Santos, I., Sene, I.G., Salomoni, S., Borges, G.A. (eds.): Development of a microcontrolled bioinstrumentation system for active control of leg prostheses. In: 2008 30th Annual International Conference of the IEEE Engineering in Medicine and Biology Society. IEEE (2008)
96. Delis, A.L., de Carvalho, J.L.A., Borges, G.A., de Siqueira Rodrigues, S., dos Santos, I., da Rocha, A.F. (eds.): Fusion of electromyographic signals with proprioceptive sensor data in myoelectric pattern recognition for control of active transfemoral leg prostheses. In: 2009 Annual International Conference of the IEEE Engineering in Medicine and Biology Society. IEEE (2009)
97. Delis, A.L., de Carvalho, J.L.A., da Rocha, A.F., de Oliveira Nascimento, F.A., de Araújo Borges, G. (eds.): Development of a myoelectric controller based on knee angle estimation. Biodevices (2009)
98. Delis, A.L., De Carvalho, J.L.A., Seisdedos, C.V., Borges, G.A., Da Rocha, A.F. (eds.): Myoelectric control algorithms for leg prostheses based on data fusion with proprioceptive sensors. In: Proceedings ISSNIP Biosignals and Biorobotics Conference (2010)
99. Scandaroli, G.G., Borges, G.A., Ishihara, J.Y., Terra, M.H., da Rocha, A.F., de Oliveira Nascimento, F.A. (eds.): Estimation of foot orientation with respect to ground for an above knee robotic prosthesis. In: 2009 IEEE/RSJ International Conference on Intelligent Robots and Systems. IEEE (2009)
100. Farmer, S., Silver-Thorn, B., Voglewede, P., Beardsley, S.A.: Within-socket myoelectric prediction of continuous ankle kinematics for control of a powered transtibial prosthesis. J. Neural Eng. **11**(5), 056027 (2014)
101. Silver-Thorn, B., Current, T., Kuhse, B.: Preliminary investigation of residual limb plantarflexion and dorsiflexion muscle activity during treadmill walking for trans-tibial amputees. Prosthet. Orthot. Int. **36**(4), 435–442 (2012)
102. Park, T.-G., Kim, J.-Y.: Real-time prediction of walking state and percent of gait cycle for robotic prosthetic leg using artificial neural network. Intel. Serv. Robot. **15**(4), 527–536 (2022)
103. Eslamy, M., Alipour, K. (eds.): Design of a high level controller for active foot prostheses using Gaussian process intent recognition. In: 2017 5th RSI International Conference on Robotics and Mechatronics (ICRoM). IEEE (2017)
104. Eslamy, M., Schilling, A.F. (eds.): A conceptual high level controller to walk with active foot prostheses/orthoses. In: 2018 7th IEEE International Conference on Biomedical Robotics and Biomechatronics (Biorob). IEEE (2018)
105. Eslamy, M., Oswald, F., Schilling, A.F.: Estimation of knee angles based on thigh motion: a functional approach and implications for high-level controlling of active prosthetic knees. IEEE Control Syst. Mag. **40**(3), 49–61 (2020)
106. Eslamy, M., Oswald, F., Schilling, A.F. (eds.): Motion planning for active prosthetic knees. In: 2020 8th IEEE RAS/EMBS International Conference for Biomedical Robotics and Biomechatronics (BioRob). IEEE (2020)
107. Dey, S., Eslamy, M., Yoshida, T., Ernst, M., Schmalz, T., Schilling, A. (eds.): A support vector regression approach for continuous prediction of ankle angle and moment during walking: an implication for developing a control strategy for active ankle prostheses. In: 2019 IEEE 16th International Conference on Rehabilitation Robotics (ICORR), 24–28 June 2019. https://doi.org/10.1109/ICORR.2019.8779445
108. Dey, S., Yoshida, T., Ernst, M., Schmalz, T., Schilling, A.F. (eds.): A random forest approach for continuous prediction of joint angles and moments during walking: an implication for controlling active knee-ankle prostheses/orthoses. In: 2019 IEEE International Conference on Cyborg and Bionic Systems (CBS). IEEE (2019)

109. Dey, S., Yoshida, T., Foerster, R.H., Ernst, M., Schmalz, T., Schilling, A.F.: Continuous prediction of joint angular positions and moments: a potential control strategy for active knee-ankle prostheses. IEEE Trans. Med. Robot. Bionics **3**, 347–355 (2020)
110. Chen, J., Zhang, X., Cheng, Y., Xi, N.: Surface EMG based continuous estimation of human lower limb joint angles by using deep belief networks. Biomed. Signal Process. Control **40**, 335–342 (2018)
111. Boudali, A.M., Sinclair, P.J., Manchester, I.R. (eds.): Prediction of smooth gait transitioning for active lower limb prosthetics. In: 2019 41st Annual International Conference of the IEEE Engineering in Medicine and Biology Society (EMBC). IEEE (2019)
112. Rai, V., Rombokas, E. (eds.): A framework for mode-free prosthetic control for unstructured terrains. In: 2019 IEEE 16th International Conference on Rehabilitation Robotics (ICORR). IEEE (2019)
113. Rai, V., Sharma, A., Preechayasomboon, P., Rombokas, E. (eds.): Coordinated movement for prosthesis reference trajectory generation: temporal factors and attention. In: 2020 8th IEEE RAS/EMBS International Conference for Biomedical Robotics and Biomechatronics (BioRob). IEEE (2020)
114. Rai, V., Sharma, A., Rombokas, E. (eds.): Mode-free control of prosthetic lower limbs. In: 2019 International Symposium on Medical Robotics (ISMR). IEEE (2019)
115. Sharma, A., Rombokas, E.: Improving IMU-based prediction of lower limb kinematics in natural environments using egocentric optical flow. IEEE Trans. Neural Syst. Rehabil. Eng. **30**, 699–708 (2022)
116. Gupta, R., Dhindsa, I.S., Agarwal, R.: Continuous angular position estimation of human ankle during unconstrained locomotion. Biomed. Signal Process. Control **60**, 101968 (2020)
117. Fereydooni, R.H., Siahkali, H., Shayanfar, H.A., Mazinan, A.H.: sEMG-based variable impedance control of lower-limb rehabilitation robot using wavelet neural network and model reference adaptive control. Indus. Robot: Int. J. Robot. Res. Appl. (2020)
118. Keleş, A.D., Yucesoy, C.A.: Development of a neural network based control algorithm for powered ankle prosthesis. J. Biomech. **113**, 110087 (2020)
119. Bovi, G., Rabuffetti, M., Mazzoleni, P., Ferrarin, M.: A multiple-task gait analysis approach: kinematic, kinetic and EMG reference data for healthy young and adult subjects. Gait Post. **33**(1), 6–13 (2011). https://doi.org/10.1016/j.gaitpost.2010.08.009
120. Al Kouzbary, H., Al Kouzbary, M., Tham, L.K., Liu, J., Shasmin, H.N., Osman, N.A.A.: Generating an adaptive and robust walking pattern for a prosthetic ankle-foot by utilizing a nonlinear autoregressive network with exogenous inputs. IEEE Trans. Neural Netw. Learn. Syst. (2021)
121. Al Kouzbary, M., Abu Osman, N.A., Al Kouzbary, H., Shasmin, H.N., Arifin, N.: Towards universal control system for powered ankle–foot prosthesis: a simulation study. Int. J. Fuzzy Syst. **22**, 1299–1313 (2020)
122. Li, M., Liu, W., Si, J., Stallrich, J.W., Huang, H.: Hierarchical optimization for control of robotic knee prostheses toward improved symmetry of propulsive impulse. IEEE Trans. Biomed. Eng. **70**(5), 1634–1642 (2023)
123. Wu, H., Dyson, M., Nazarpour, K.: Internet of Things for beyond-the-laboratory prosthetics research. Philos. Trans. Royal Soc. A: Math., Phys. Eng. Sci. (2022). https://doi.org/10.1098/rsta.2021.0005
124. Perez, A.J., Siddiqui, F., Zeadally, S., Lane, D.: A review of IoT systems to enable independence for the elderly and disabled individuals. Internet of Things (2023). https://doi.org/10.1016/j.iot.2022.100653
125. Badugu, S., Srikanth, K., Inampudi, L.N.: IoT for healthcare. Int. J. Sci. Res. (IJSR) **5**, 2319–7064 (2016)
126. Khan, M., Han, K., Karthik, S.: Designing smart control systems based on Internet of Things and big data analytics. Wirel. Pers. Commun. **99**(4), 1683–1697 (2018)
127. Azzawi, M., Hassan, R., Abu Bakar, K.A.: A review on Internet of Things (IoT) in healthcare. Int. J. Appl. Eng. Res. **11**, 10216–10221 (2016)

128. Jagadeeswari, V., Subramaniyaswamy, V., Logesh, R., Vijayakumar, V.: A study on medical Internet of Things and big data in personalized healthcare system. Health Inform. Sci. Syst. **6**(1), 14 (2018)
129. Othman, A., Elsheikh, A., Al-Mutawaa, A.M. (eds.): Internet of Things (IoT) for elderly's healthcare and wellbeing: applications, prospects and challenges. In: 2023 IEEE Symposium on Computers and Communications (ISCC), pp. 9–12 (2023). https://doi.org/10.1109/ISCC58397.2023.10218131
130. Farahani, B., Firouzi, F., Chakrabarty, K.: Healthcare IoT. In: Firouzi, F., Chakrabarty, K., Nassif, S. (eds.) Intelligent Internet of Things: From Device to Fog and Cloud, pp. 515–545. Springer International Publishing, Cham (2020)
131. Prabavathy, S., Reddy, R.: Exploring the Dichotomy on Opportunities and Challenges of Smart Technologies in Healthcare Systems, pp. 1–24 (2022)
132. Pradhan, B., Bharti, D., Chakravarty, S., Ray, S.S., Voinova, V.V., Bonartsev, A.P., Pal, K.: Internet of Things and robotics in transforming current-day healthcare services. J. Healthc. Eng. **2021**, 9999504 (2021). https://doi.org/10.1155/2021/9999504
133. Peng, H., Tian, Y., Kurths, J., Li, L., Yang, Y., Wang, D.: Secure and energy-efficient data transmission system based on chaotic compressive sensing in body-to-body networks. IEEE Trans. Biomed. Circ. Syst. **11**(3), 558–573 (2017). https://doi.org/10.1109/TBCAS.2017.2665659
134. Pradhan, B., Bhattacharyya, S., Pal, K.: IoT-based applications in healthcare devices. J. Healthc. Eng. (2021). https://doi.org/10.1155/2021/6632599
135. Adinolfi, P., Starace, F., Palumbo, R.: Health outcomes and patient empowerment: the case of health budgets in Italy. J. Health Manag. **18**(1), 117–133 (2016). https://doi.org/10.1177/0972063415625524
136. Toni, M., Mattia, G.: The evolution to patient centricity. In: Toni, M., Mattia, G. (eds.) The Digital Healthcare Revolution: Towards Patient Centricity with Digitization, Service Innovation and Value Co-creation, pp. 51–62. Springer International Publishing, Cham (2022)
137. Beyrouthy, T., Kork, S.K.A., Korbane, J.A., Abdulmonem, A. (eds.): EEG mind controlled smart prosthetic arm. In: 2016 IEEE International Conference on Emerging Technologies and Innovative Business Practices for the Transformation of Societies (EmergiTech), 3–6 August 2016. https://doi.org/10.1109/EmergiTech.2016.7737375
138. Williams, R.J., Holloway, C., Miodownik, M.: The ultimate wearable: connecting prosthetic limbs to the IoPH. In: Proceedings of the 2016 ACM International Joint Conference on Pervasive and Ubiquitous Computing: Adjunct; Heidelberg, Germany: Association for Computing Machinery, pp. 1079–1083 (2016). https://doi.org/10.1145/2968219.2972711
139. Huang, G.S., Chang, S.C., Chen, C.C., Lai, C.L. (eds.): Development of lower limb exoskeleton robot. In: 2020 12th International Symposium on Communication Systems, Networks and Digital Signal Processing (CSNDSP), 20–22 July 2020. https://doi.org/10.1109/CSNDSP49049.2020.9249514
140. Silva, R., Starliper, N., Zhong, B., Huang, H., Lobaton, E.: Evaluation of embedded platforms for lower limb prosthesis with visual sensing capabilities (2020)
141. LeMoyne, R.: Future and advanced concepts for the powered prosthesis. In: LeMoyne, R. (ed.): Advances for Prosthetic Technology: From Historical Perspective to Current Status to Future Application, pp. 127–130. Springer, Japan, Tokyo (2016)
142. Clemens, S., Doerger, C., Lee, S.-P.: Current and emerging trends in the management of fall risk in people with lower limb amputation. Curr. Geriatr. Rep. **9**(3), 134–141 (2020)
143. Heinemann, A.W., Ehrlich-Jones, L., Connelly, L., Semik, P., Fatone, S.: Enhancing quality of prosthetic services with process and outcome information. Prosthet. Orthot. Int. **41**(2), 164–170 (2016). https://doi.org/10.1177/0309364616637957
144. Stepien, J., Cavenett, S., Taylor, L., Crotty, M.: Activity levels among lower-limb amputees: self-report versus step activity monitor. Arch. Phys. Med. Rehabil. **88**, 896–900 (2007)
145. Solhan, M.B., Trull, T.J., Jahng, S., Wood, P.K.: Clinical assessment of affective instability: comparing EMA indices, questionnaire reports, and retrospective recall. Psychol. Assess. **21**(3), 425–436 (2009)

146. Ramstrand, N., Brodtkorb, T.-H.: Considerations for developing an evidenced-based practice in orthotics and prosthetics. Prosthet. Orthot. Int. **32**(1), 93–102 (2008)
147. Chadwell, A., Diment, L., Micó-Amigo, M., Morgado Ramírez, D.Z., Dickinson, A., Granat, M., Kenney, L., Kheng, S., Sobuh, M., Ssekitoleko, R., Worsley, P.: Technology for monitoring everyday prosthesis use: a systematic review. J. Neuroeng. Rehabil. **17**(1), 93 (2020). https://doi.org/10.1186/s12984-020-00711-4
148. Hafner, B.J., Sanders, J.E.: Considerations for development of sensing and monitoring tools to facilitate treatment and care of persons with lower-limb loss: a review. J. Rehabil. Res. Dev. **51**(1), 1–14 (2014). https://doi.org/10.1682/JRRD.2013.01.0024
149. Balkman, G.S., Vamos, A.C., Sanders, J.E., Larsen, B.G., Hafner, B.J.: Prosthetists' perceptions of information obtained from a lower limb prosthesis monitoring system: a pilot study. J. Prosthet. Orthot. JPO **31**(2), 112–120 (2019). https://doi.org/10.1097/JPO.0000000000000203
150. Mamdiwar, S.D., Akshith, R., Shakruwala, Z., Chadha, U., Srinivasan, K., Chang, C.-Y.: Recent advances on IoT-assisted wearable sensor systems for healthcare monitoring. Biosensors [Internet]. **11**(10) (2021). https://doi.org/10.3390/bios11100372
151. Alwan, M., Sifferlin, E.B., Turner, B., Kell, S., Brower, P., Mack, D.C., Dalal, S., Felder, R.A.: Impact of passive health status monitoring to care providers and payers in assisted living. Telemed. e-Health **13**(3), 279–285 (2007). https://doi.org/10.1089/tmj.2006.0056
152. Gardner, D.W., Redd, C.B., Cagle, J.C., Hafner, B.J., Sanders, J.E.: Monitoring prosthesis user activity and doffing using an activity monitor and proximity sensors. JPO: J. Prosthet. Orthot. **28**(2) (2016)
153. Maqbool, H.F., Husman, M.A.B., Awad, M.I., Abouhossein, A., Iqbal, N., Dehghani-Sanij, A.A.: A real-time gait event detection for lower limb prosthesis control and evaluation. IEEE Trans. Neural Syst. Rehabil. Eng. **25**(9), 1500–1509 (2017). https://doi.org/10.1109/TNSRE.2016.2636367
154. Appelboom, G., Camacho, E., Abraham, M.E., Bruce, S.S., Dumont, E.L., Zacharia, B.E., D'Amico, R., Slomian, J., Reginster, J.Y., Bruyère, O.: Smart wearable body sensors for patient self-assessment and monitoring. Arch. Public Health **72**(1), 1–9 (2014)
155. Mansour, H.A., Abdullah, H.N. (eds.): Design and implementation of remote evaluation system for the prosthetic limbs. In: 2022 3rd International Conference for Emerging Technology (INCET), 27–29 May 2022 (2022). https://doi.org/10.1109/INCET54531.2022.9824666
156. Rattanakoch, J., Samala, M., Limroongreungrat, W., Guerra, G., Tharawadeepimuk, K., Nanbancha, A., Niamsang, W., Kerdsomnuek, P., Suwanmana, S.: Validity and reliability of inertial measurement unit (IMU)-derived 3D joint kinematics in persons wearing transtibial prosthesis. Sensors [Internet] **23**(3) (2023). https://doi.org/10.3390/s23031738
157. Tschiedel, M., Russold, M.F., Kaniusas, E., Vincze, M.: Real-time limb tracking in single depth images based on circle matching and line fitting. Vis. Comput. **38**(8), 2635–2645 (2022)
158. de Miguel-Fernández, J., Lobo-Prat, J., Prinsen, E., Font-Llagunes, J.M., Marchal-Crespo, L.: Control strategies used in lower limb exoskeletons for gait rehabilitation after brain injury: a systematic review and analysis of clinical effectiveness. J. Neuroeng. Rehabil. **20**(1), 23 (2023). https://doi.org/10.1186/s12984-023-01144-5
159. Halim, A., Ibrahim, A.A., Awad, M.I., Atia, M.R.A. (eds.): Optimization of sensor number for lower limb prosthetics using genetic algorithm. In: 2020 International Conference on Innovative Trends in Communication and Computer Engineering (ITCE), 8–9 February 2020 (2020). https://doi.org/10.1109/ITCE48509.2020.9047780
160. Bryant, M.: How AI and machine learning are changing prosthetics: MedTech Dive (2019). https://www.medtechdive.com/news/how-ai-and-machine-learning-are-changing-prosthetics/550788/
161. Paternò, L., Dhokia, V., Menciassi, A., Bilzon, J., Seminati, E.: A personalised prosthetic liner with embedded sensor technology: a case study. Biomed. Eng. Online **19**(1), 71 (2020). https://doi.org/10.1186/s12938-020-00814-y

162. Atzori, L., Iera, A., Morabito, G.: The Internet of Things: a survey. Comput. Netw. **54**(15), 2787–2805 (2010). https://doi.org/10.1016/j.comnet.2010.05.010
163. Diffie, W., Hellman, M.: New directions in cryptography. IEEE Trans. Inf. Theory **22**(6), 644–654 (1976). https://doi.org/10.1109/TIT.1976.1055638
164. Pană, C.F., Manta, L.F., Vladu, I.C., Cismaru, Ş.I., Petcu, F.L., Cojocaru, D., Bîzdoacă, N.: The design of a smart lower-limb prosthesis supporting people with transtibial amputation—a data acquisition system. Appl. Sci. [Internet] **12**(13) (2022). https://doi.org/10.3390/app12136722
165. Morales, J.: What is Zigbee and how does it work? MUO (2022). https://www.makeuseof.com/what-is-zigbee/

Application of Model-Agnostic Meta Learning (MAML) in Automated Invasive Ductal Carcinoma Grading

Yan Chai Hum, Wingates Voon, and Khin Wee Lai

1 Breast Cancer: Prevalence, Prognosis, and Advances in Automated Grading

In 2020, approximately 19.3 million new cancer cases were diagnosed globally, leading to nearly 10 million cancer-related deaths. Among women, breast cancer has emerged as the most prevalent type, with an estimated 2.3 million new cases reported annually [46]. Breast cancer involves a spectrum of disorders characterized by the uncontrolled multiplication of breast cells, resulting in the formation of a lump in a localized area of the breast [2]. Invasive ductal carcinoma (IDC) is the most common form, accounting for over 80% of breast cancer cases [40]. Early detection and screening are essential in effectively combating breast cancer. Standard screening methods include mammography, magnetic resonance imaging (MRI), and ultrasound imaging of the breast. If suspicious tissue is found, a biopsy is performed for histological examination. Following tissue extraction, three preparatory steps precede histological grading: (1) formalin fixation, (2) paraffin section embedment, and (3) hematoxylin and eosin staining [34].

Three primary prognostic markers guide the treatment of breast cancer: (1) lymph node (LN) status, (2) tumor size, and (3) histological grade [41]. Studies have demonstrated that the prognosis based on histological grade is as significant as LN status and even more indicative than tumor size [17, 47]. Furthermore, the accuracy of clinical outcome predictions improves when histological grade and LN status are considered together [21]. For instance, [16] found that histological grade accurately

Y. C. Hum (✉) · W. Voon
Department of Mechatronics and Biomedical Engineering, Lee Kong Chian Faculty of Engineering and Science, Universiti Tunku Abdul Rahman, Sungai Long, Malaysia
e-mail: humyc@utar.edu.my

K. W. Lai
Department of Biomedical Engineering, Universiti Malaya, Kuala Lumpur, Malaysia
e-mail: lai.khinwee@um.edu.my

K. W. Lai and P. S. Q. Yeoh (eds.), *Biomedical Engineering*, Series in BioEngineering,
https://doi.org/10.1007/978-981-97-9294-8_7

predicted tumor behavior, particularly in early-stage small tumors. Similarly, [37] noted that patients with high-grade breast cancer who underwent mastectomy experienced higher mortality rates and greater axillary lymph node involvement than those with lower-grade tumors. Consequently, the breast cancer grade, specifically the IDC grade, serves as a critical indicator of patient outcomes.

The grade of breast cancer reflects the tumor's aggressiveness [26]. Pathologists use the Nottingham Grading Scheme (NGS) to categorize breast cancer into grades based on three morphological traits: (1) mitotic count (number of proliferating tumor cells), (2) nuclear pleomorphism (overall appearance of tumor cells), and (3) degree of tubule formation (how closely tumor cells replicate normal glands) [34]. These traits are combined to produce a total score that classifies breast cancer as low-grade (grade 1), intermediate-grade (grade 2), or high-grade (grade 3) [26]. Despite being the gold standard for cancer diagnosis, manual breast cancer grading is labor-intensive, time-consuming, and subjective, with pathologist competence significantly impacting results [20]. Inexperienced pathologists may make incorrect diagnoses [7]. Elmore et al. [14] reported that there was an overall agreement of approximately 75.3% between pathologists' assessments and expert consensus-derived reference diagnoses. Additionally, manual grading of low-magnification photographs is prone to statistical, distributional, and human errors [23].

As computer vision technology has advanced, automated breast cancer grading methods have gained popularity. Previous research [6, 10, 13, 31] attempted to enhance the manual breast cancer grading process by integrating NGS criteria with traditional machine learning approaches. However, these methods are often highly feature-dependent, time-consuming, and computationally expensive. Wan et al. [53] introduced a new approach using deep learning by employing a Convolutional Neural Network (CNN) to classify breast cancer grades. Subsequently, several studies [27, 38, 56, 57] have utilized various deep learning techniques to address this classification challenge. Although these techniques are robust, they require substantial computational resources. In contrast, transfer learning has become more prevalent, many studies [1, 58] have applied transfer learning to breast cancer grading. Notably, there is a knowledge gap in the literature: no studies have comprehensively compared the performance of recent pre-trained state-of-the-art CNN architectures such as EfficientNetB0 [48], EfficientNetV2B0 [49], EfficientNetV2B0-21k [49], ResNetV1-50 [18], ResNetV2-50 [19], MobileNetV1 [22], and MobileNetV2 [36]. As a result, there is limited awareness of how CNN structures are employed in automatic IDC grading. This research aims to fill this gap by presenting findings on the application of various CNN architectures in automated IDC grading, ranging from simple, lightweight models to complex, heavyweight models.

As the field of breast cancer research continues to evolve, the adoption of sophisticated computational techniques has become essential for advancing diagnostic accuracy and efficiency. The application of deep learning has particularly transformed the landscape, offering powerful tools to enhance the detection and classification of invasive ductal carcinoma (IDC). However, the resource-intensive nature of deep learning models poses significant challenges, prompting researchers to explore alternative approaches such as transfer learning. By leveraging pre-trained models on large datasets, transfer learning enables more efficient learning, reducing the need for extensive computational resources and large annotated datasets. Despite the strides made in these areas, there remains a critical need to evaluate and compare the effectiveness of cutting-edge CNN architectures in the context of IDC grading. This exploration is vital to unlocking the full potential of deep learning and ensuring that these technologies are accessible and applicable across diverse clinical settings. As we delve into the advances in automated IDC grading systems, it becomes imperative to examine the historical context and foundational methodologies that have paved the way for contemporary innovations.

2 Advances in Automated IDC Grading Systems

Early automated systems for invasive ductal carcinoma (IDC) primarily aimed to extract distinct features to characterize tumor morphology and texture. These systems leveraged a variety of methods, including multi-field-of-view (multi-FOV) classifiers [6], Grassmann manifold-based vectors of locally aggregated descriptors (VLAD) representations [12], and multi-level feature sets with graph embedding [53]. The multi-FOV classifiers aimed to provide a comprehensive view of the tumor by examining it from multiple perspectives, thereby capturing intricate details of its texture and structure. Grassmann manifold-based VLAD representations offered a mathematical approach to managing the complex variations in tumor morphology by aggregating local descriptors into a compact and informative representation. Meanwhile, the use of multi-level feature sets with graph embedding allowed the modeling of complex relationships within the histological data, highlighting the interactions between various cellular structures. Despite their sophistication, these methods often lacked effective heuristics for feature extraction, which limited their adaptability and accuracy. This gap prompted the shift towards more advanced deep learning-based techniques, which promised better performance and efficiency.

Deep learning, particularly Convolutional Neural Networks (CNNs), has emerged as the gold standard for histological image classification due to its robust capability to automatically extract relevant features from raw data [28]. CNNs can learn complex patterns in histological images that are difficult to capture through manual feature extraction, thereby enhancing the precision of IDC classification. Recent advancements in this domain include the development of an entropy-based elastic ensemble (3E-Net) [38], which dynamically combines predictions from multiple models to improve accuracy and robustness. The use of DenseNets and ResNets ensemble models [5] further enhances the learning process by leveraging deep residual learning and dense connections, facilitating the flow of information across layers and improving model performance on complex histopathological tasks.

Additionally, nuclei-guided models (NGNet) [56] have been introduced to focus on the cellular structures within the tissue, providing a more targeted approach to identifying malignancies. Hybrid deep learning classifiers [35] combine the strengths of different neural network architectures to create a more versatile and powerful model for IDC classification. Transfer learning has also played a significant role, with pre-trained CNNs from ImageNet being widely employed to improve IDC grading accuracy [51, 52]. This approach leverages existing knowledge from large-scale datasets, reducing the need for extensive labeled histopathological data and accelerating model development.

However, these advanced techniques often depend on the availability of large, annotated datasets to achieve optimal performance [25]. The annotation process is labor-intensive and requires expert knowledge, posing a challenge for the widespread implementation of these methods in clinical settings. As a result, there is a growing interest in developing more efficient algorithms that can perform well with limited data, potentially utilizing semi-supervised or unsupervised learning strategies to overcome these limitations and broaden the applicability of deep learning in IDC diagnosis.

The reliance on large, annotated datasets presents a significant hurdle for the deployment of advanced deep learning techniques in practical clinical settings. The process of data annotation is not only resource-intensive but also requires a high degree of expertise, which can be a bottleneck in scaling these technologies for widespread use. As the demand for efficient diagnostic tools grows, there is an urgent need to explore alternative methodologies that can deliver robust performance with minimal data input. This necessity has driven interest in innovative algorithms capable of functioning effectively in data-scarce environments. Among the most promising of these approaches are semi-supervised and unsupervised learning strategies, which offer the potential to harness unlabeled data and extract meaningful insights with reduced reliance on extensive labeled datasets. As we transition into exploring few-shot learning (FSL) approaches in histopathology, it is important to consider how these methodologies can revolutionize the field by addressing the challenges of data scarcity and opening new avenues for artificial intelligence applications in healthcare.

3 Few-Shot Learning Approaches in Histopathology: Innovations and Applications

Few-Shot Learning (FSL) is an exciting area within the field of machine learning that focuses on teaching models to learn from a small number of examples. Traditional machine learning algorithms typically require large datasets with many labeled examples to achieve high performance [25, 28]. However, in many real-world situations, such as medical imaging or rare species identification, gathering extensive labeled data can be challenging, expensive, or time-consuming [2, 37]. FSL addresses this limitation by enabling models to generalize from just a few examples of each class. The primary objective of FSL is to mimic human-like learning abilities, where we can recognize new objects or concepts after seeing only a handful of examples [33]. This approach has opened up new possibilities for applying machine learning in scenarios where data is scarce, thus expanding the potential applications of artificial intelligence.

Few-Shot Learning (FSL) aims to classify novel classes using only a limited number of labeled samples. This approach addresses the challenge of limited data availability, which is particularly relevant in fields like medical imaging, where obtaining large datasets can be difficult. FSL techniques can be generally categorized into three main groups. The first category is metric-based methods, such as ProtoNet [43] and RelationNet [44], which focus on learning a similarity measure to compare different classes effectively. The second category includes optimization-based methods, such as MAML [15] and ANIL [33], which aim to optimize the model parameters for rapid adaptation to new tasks. The third category is hallucination-based methods, including DAGAN [3] and the meta-learner with hallucinator [54], which generate synthetic data to augment the training set and improve classification performance.

FSL has gained significant popularity in the field of histopathology due to the scarcity of annotated data, which is a common issue in this domain. For instance, Deuschel et al. [11] implemented ProtoNet to classify novel histopathological classes using minimal samples, demonstrating the potential of FSL in reducing the need for extensive labeled datasets. Singh et al. [42] utilized the Reptile algorithm [32], along with advanced data augmentations, to enhance performance in various medical classification tasks. Similarly, Shakeri et al. [39] evaluated the performance of multiple FSL methods on histopathology classification across different domain shifts, highlighting the adaptability of FSL techniques in diverse settings. Additionally, Dai et al. [9] proposed PFEMed, a method employing a dual-encoder architecture for a variational autoencoder, which enhances the model's ability to learn from few samples. Lu et al. [29] developed MedOptMed, which utilizes convex optimization classifiers tailored for medical classification tasks, offering improved accuracy and robustness. Furthermore, Maia et al. [30] compared the performance of transformers and FSL methods in classifying oral cancer histopathological images, providing insights into the strengths and weaknesses of different approaches. Carloni and Colantonio [8] proposed a CNN with a causality-factors extractor module specifically designed for few-shot breast cancer detection, showcasing the potential of FSL in improving diagnostic accuracy with limited data. Despite these advancements, the application of FSL in IDC grading remains largely unexplored, presenting an opportunity for further research and innovation in this area. To address this gap, we implemented a range of FSL methods to compare their performance in the few-shot IDC grading task, aiming to identify the most effective strategies for this specific application.

Among the optimization-based methods, MAML stands out due to its ability to achieve a robust initialization with just a few gradient descent steps through its dual optimization loops. However, the bi-optimization nature of MAML necessitates second-order computations, which can significantly increase computational costs. Additionally, the requirement to backpropagate through all feature layers further exacerbates these demands, leading to suboptimal efficiency in practical applications. Although several studies on MAML [4, 24, 55, 55] have proposed various methods to enhance computational efficiency, they have largely overlooked the potential benefits of leveraging widely available ImageNet pre-trained weights. These pre-trained weights can be frozen to capitalize on existing prior knowledge, thereby reducing the computational burden and accelerating the training process. Therefore, this study explores the integration of ImageNet pre-trained weights, along with other strategies, to improve the computational efficiency and overall effectiveness of MAML in few-shot learning scenarios. By doing so, we aim to provide a more practical and scalable solution for FSL applications, particularly in the context of IDC grading.

As the exploration of Few-Shot Learning (FSL) progresses, optimization-based methods have emerged as powerful tools, with Model-Agnostic Meta-Learning (MAML) at the forefront. MAML is particularly noted for its capability to achieve a robust initialization through its unique dual optimization loops, allowing models to rapidly adapt to new tasks with minimal data. Despite its advantages, MAML's reliance on bi-optimization and the necessity for second-order computations pose significant computational challenges, often hindering its practical application. The computational demands are further intensified by the need to backpropagate through all feature layers, which can lead to inefficiencies. Although recent studies have introduced methods to enhance MAML's computational efficiency, the potential of integrating pre-trained weights from resources like ImageNet remains underexplored. By freezing these pre-trained weights, it is possible to leverage existing knowledge, thereby reducing the computational load and accelerating the training process. This study aims to integrate these strategies to optimize MAML for few-shot learning, particularly in the context of IDC grading. Through this approach, we seek to offer a more practical and scalable solution, enhancing the applicability of FSL in real-world medical diagnostics. As we transition into discussing the specifics of MAML, it is important to understand its foundational principles and how they can be applied to improve learning efficiency in environments characterized by limited data availability.

4 Meta-Learning with Model-Agnostic Meta-Learning (MAML)

The goal of the MAML [15] is to learn the initial model parameters in such a way that the base model can quickly adapt to new unseen tasks through gradient descent. MAML consists of two optimization loops:

1. **Outer optimization loop**: This loop updates the base model parameters θ to facilitate rapid adaptation to new tasks.
2. **Inner optimization loop**: This loop adapts the model by taking the outer loop's initial parameters θ and performing a few gradient descent steps over a few examples for each task.

The base model is defined as a CNN with meta-initialization parameters θ, represented by f_θ, where $\theta = (\theta_1, ..., \theta_l)$ denotes the parameters for the l layers of the model. For any batch B of tasks $\{T_1, T_2, \ldots, T_B\}$ drawn independently from distribution $p(T)$, each task T_b is associated with a training dataset (support set) $D^{tr}_{(T_b)} = \{(x_i, y_i)\}_{i=1}^{M}$ which is used for inner loop updates, and a testing dataset (query set) $D^{ts}_{(T_b)} = \{(\widehat{x_i}, \widehat{y_i})\}_{i=1}^{\widehat{M}}$ is used for outer loop updates. Let $\theta^{Q}_{(T_b)}$ to be θ after Q gradient descents for task T_b, starting from $\theta_0 = \theta$. Thus, each update in the inner loop is computed as:

$$\theta^{Q}_{(T_b)} = \theta^{0}_{(T_b)} - \alpha \sum_{q=1}^{Q-1} \nabla_{\theta^{(q-1)}_{(T_b)}} \mathcal{L}_{D^{tr}_{(T_b)}}(f_{\theta^{(q-1)}_{(T_b)}})$$

where $\mathcal{L}_{D^{tr}_{(T_b)}}$ is the loss function on the support set of task T_b. The $\mathcal{L}_{D^{tr}_{(T_b)}}$ measures the discrepancy between the model predictions $f_\theta(x_i)$ and ground truth y_i. For classification tasks, we use the cross-entropy loss, which is defined as:

$$\mathcal{L}(f_\theta) = \sum_{x_i, y_i} y_i \log(f_\theta(x_i)) + (1 - y_i)\log(1 - f_\theta(x_i))$$

Following the inner optimization loop, MAML calculates and accumulates the query losses to update θ, resulting in a new meta-initialization θ. The outer optimization loop is described as:

$$\theta = \theta - \beta \nabla_\theta \sum_{b=1}^{B} \mathcal{L}_{D^{ts}_{(T_b)}}(f_{\theta^{Q}_{(T_b)}})$$

where θ represents the meta-parameters, β is the learning rate for the outer optimization loop. $\mathcal{L}_{D^{ts}_{(T_b)}}$ is the loss on the query set of tasks T_b after Q inner loop updates. During the meta-testing phase, novel unseen tasks T_b sampled from D_{novel} are drawn from the task distribution $p(T)$, and the loss and accuracy are evaluated on $D^{ts}_{(T_b)}$ after inner loop adaptation using $D^{tr}_{(T_b)}$. The training details of MAML are demonstrated in Algorithm 1. In the implementation of First-order MAML (FOMAML), the second-order term in the MAML update is omitted to enhance computational efficiency, although this approach sacrifices some second-order information.

Algorithm 1: MAML for Supervised Few-shot IDC Grading

Result: Trained model
Required: $p(T)$: distribution over tasks
Required: α, β : inner and outer learning rate hyperparameters
Required: E_{total} : number of total epochs
Required: B : number of tasks per batch
Required: $M, \hat{M}$: number of data points for $D^{tr}_{(T_b)}$ and $D^{ts}_{(T_b)}$
Required: Q : number of inner adaptation steps
Randomly initialize meta-parameters θ
for e *in* E_{total} **do**
  Sample B number of tasks from $p(T)$
  for $T_b = T_1, T_2, ..., T_B$ **do**
    Sample M data points $D^{tr}_{(T_b)} = \{x_i, y_i\}$ from T_b
    for $q = 1, 2, ..., Q$ **do**
      Evaluate $\nabla_{\theta^{(q)}_{(T_b)}}$ using $D^{tr}_{(T_b)}$ and $L_{D^{tr}_{(T_b)}}$
      Compute adapted parameters with gradient descent:
      $\theta^{(q)}_{(T_b)} = \theta^{(q-1)}_{(T_b)} - \alpha \nabla_{\theta^{(q-1)}_{(T_b)}} L_{D^{tr}_{(T_b)}}(f_{\theta^{(q-1)}_{(T_b)}})$
    end
    Sample $\hat{M}$ data points $D^{ts}_{(T_b)} = \{\hat{x}_i, \hat{y}_i\}$ from T_b for meta-update
  end
  Update $\theta \leftarrow \theta - \beta \nabla_\theta \sum_{b=1}^{B} L_{D^{ts}_{(T_b)}}(f_{\theta^{Q}_{(T_b)}})$
  if *early stopping criterion is achieved* **then**
    break
  end
end

The application of Model-Agnostic Meta Learning (MAML) within the context of automated IDC grading illuminates its broader utility in addressing rare forms of breast cancer as well. As showcased in Table 1, breast cancer manifests in various forms beyond the prevalent IDC, such as Lobular Carcinoma, Papillary Carcinoma, and Mucinous Carcinoma. Each of these rarer cancer types presents unique challenges due to the scarcity of extensive datasets, making the adaptability of MAML particularly beneficial.

MAML's ability to fine-tune models quickly with just a few examples from each new task allows it to effectively adapt to less common forms of breast cancer, which are typically underrepresented in training data. This adaptability is crucial for developing diagnostic tools that are both inclusive and comprehensive in their ability to diagnose a wide range of breast cancer types. By training a base model on a common type like IDC and then quickly adapting to other rare types, MAML can help bridge the data gap and improve diagnostic accuracy across less common cancer classes.

The advantage of MAML in these applications lies in its two-level optimization process, which fine-tunes the model to specific tasks with minimal data input. This characteristic is essential when dealing with rare cancer types where collecting large datasets is often unfeasible. The method's flexibility and efficiency make it a potent tool for medical researchers and clinicians aiming to develop robust diagnostic systems that are responsive to the diverse manifestations of breast cancer. Thus, MAML not only enhances the scope of automated grading systems in common IDC cases but also extends its applicability to a broader spectrum of rare histopathological classifications, potentially leading to more personalized and precise cancer treatment strategies.

The versatility and effectiveness of Model-Agnostic Meta-Learning (MAML) are highlighted in its two-level optimization process, which enables fine-tuning of models using minimal data input. This adaptability is particularly crucial in scenarios involving rare cancer types, where gathering large datasets is often impractical. MAML's ability to quickly adapt to new and specific tasks makes it an invaluable tool for medical researchers and clinicians striving to create robust diagnostic systems capable of addressing the diverse presentations of breast cancer. By enhancing the efficiency of automated grading systems in common IDC cases and expanding its applicability to rare histopathological classifications, MAML fosters the development of more personalized and precise cancer treatment strategies. As we shift our focus to the datasets that underpin IDC research, these resources are foundational to advancing the capabilities of diagnostic models and improving clinical outcomes. These datasets not only support the development of more accurate grading systems but also facilitate the application of innovative machine learning techniques, such as MAML, in medical diagnostics.

5 Common Datasets for Invasive Ductal Carcinoma Research

Invasive ductal carcinoma (IDC) research relies heavily on high-quality histopathological image datasets to develop and validate diagnostic and classification models. These datasets provide researchers with a wide array of annotated images that are essential for training machine learning algorithms. By leveraging these datasets, researchers can improve the accuracy and efficiency of IDC diagnosis, contributing to better patient outcomes. In the following sections, we will explore some of the most widely used datasets in IDC research, including the BreaKHis dataset, the Breast Carcinoma Histological Images (BCHI) dataset, and the Pathological Image Dataset for IDC Grading (PathoIDCG). Each of these datasets offers unique features and characteristics that make them invaluable resources for advancing the field of IDC research.

5.1 The BreaKHis Dataset

The BreaKHis dataset [45] is a comprehensive collection of 7,909 histopathological breast cancer images obtained from 82 patients. This dataset is particularly valuable due to its diverse range of magnification levels, with images captured using magnifying factors of 40X, 100X, 200X, and 400X. These correspond to objective lenses of 4X, 10X, 20X, and 40X, respectively. Each image is digitally processed into an RGB format with a resolution of 700 by 460 pixels, ensuring consistency and clarity across the dataset. BreaKHis includes eight different classes of breast cancer, offering a wide spectrum of histological variations. This diversity is crucial for training models that can generalize across different types of breast cancer tissue. The dataset's standardized format and extensive labeling make it a cornerstone resource for researchers aiming to develop and refine diagnostic algorithms.

For a comprehensive visual and quantitative overview of the BreaKHis dataset, refer to Fig. 1 and Table 1. Figure 1 displays representative images from the dataset across eight distinct classes and four levels of magnification. Table 1, meanwhile, provides a detailed distribution of samples within these classes and magnifications. The table organizes breast cancer classes such as Adenosis, Phyllodes Tumor, Tubular Adenoma, Lobular Carcinoma, Papillary Carcinoma, Mucinous Carcinoma, Fibroadenoma, and Ductal Carcinoma vertically, with subdivisions based on sample counts at 4X, 10X, 20X, and 40X magnifications. Notably, Adenosis samples decrease from 114 at 4X to 106 at 40X, a pattern consistent across most classes where sample numbers reduce as magnification increases. Ductal Carcinoma, with the highest number of samples at all magnifications, underscores its prevalence or primary focus in the dataset. Each magnification level's total sample count also shows a decrease from 1995 at 4X to 1820 at 40X, illustrating how the dataset's composition changes with the level of detail provided by increasing magnification. This granular detail is vital for understanding the dataset's utility and ensuring that image analysis algorithms are effectively tailored and tested across different types of cancer and magnification levels.

5.2 BCHI Dataset

The Breast Carcinoma Histological Images (BCHI) dataset [59] provides a curated collection of IDC images stained with hematoxylin and eosin (H&E). Originating from the pathology department at "Agios Pavlos" Hospital in Thessaloniki, Greece, this dataset offers 282 unique images after removing duplicates. Each image has a high-resolution of 1280 $\times$ 960 pixels, providing detailed visual data for analysis. The BCHI dataset classifies IDC into three grades: Grade 1 with 98 images, Grade 2 with 102 images, and Grade 3 with 82 images. These grades correspond to different levels of tumor aggressiveness, offering researchers the opportunity to develop models that can distinguish between varying severity levels of IDC. The images were captured using a Nikon camera with a 40X magnification objective lens on a compound microscope, ensuring high-quality and precise imaging. Sample images from the BCHI dataset are displayed in Fig. 2.

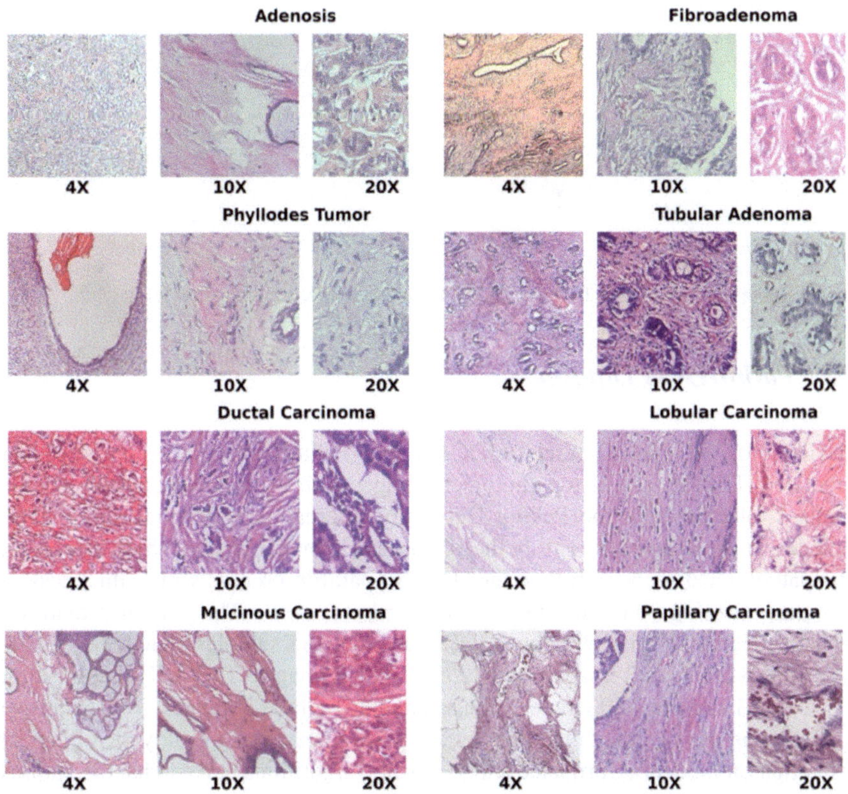

Fig. 1 Representative images from the BreaKHis dataset across eight classes and four magnification levels

Table 1 Distribution of BreaKHis samples across different classes and magnification levels

Class	Magnification			
	4X	**10X**	**20X**	**40X**
Adenosis	114	113	111	106
Phyllodes tumor	149	150	140	130
Tubular adenoma	109	121	108	115
Lobular carcinoma	156	170	163	137
Papillary carcinoma	145	142	135	138
Mucinous carcinoma	205	222	196	169
Fibroadenoma	253	260	264	237
Ductal carcinoma	864	903	896	788
Total	**1995**	**2081**	**2013**	**1820**

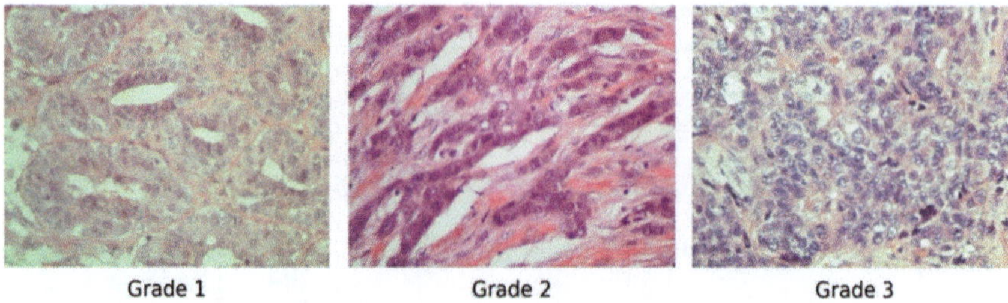

Fig. 2 Representative images from the BCHI dataset with 40X magnification level

5.3 PathoIDCG Dataset

The Pathological Image Dataset for IDC Grading (PathoIDCG) [57] comprises 3,744 annotated H&E-stained histopathological images with a resolution of 1000 × 1000 pixels. Collected from the Peking University International Hospital, this dataset includes images captured with both 20X and 40X magnification objective lenses. The pathological sections follow the standard paraffin process with a thickness of 3–5 μm, ensuring consistent preparation across samples. The dataset provides a robust collection of images for each IDC grade: the 20X samples include 600 Grade 1 images, 641 Grade 2 images, and 1,245 Grade 3 images. The 40X samples offer 317 Grade 1 images, 480 Grade 2 images, and 361 Grade 3 images. This extensive collection across two magnification levels allows researchers to study the effects of resolution on model performance and to develop algorithms that can handle various levels of detail in histopathological images. For examples of these images, refer to Fig. 3.

Building on the advancements of Model-Agnostic Meta Learning (MAML) and its impact on the field of automated IDC grading, it becomes imperative to explore other innovative methods within the realm of Few-Shot Learning (FSL) that address similar challenges of data scarcity and high variability in medical diagnostics. This leads us to the exploration of Metric-based FSL methods, which provide a robust framework for tackling the inherent difficulties of training models with limited datasets. These methods are particularly pertinent in fields like histopathology, where the availability of extensively labeled datasets is often a significant hurdle.

6 Metric-Based Few-Shot Learning (FSL) Methods

Metric-based Few-Shot Learning (FSL) methods are designed to solve the challenge of limited data by learning a similarity metric that quantifies the comparison between a small number of examples in a support set and new examples in a query set. These approaches leverage the power of learning from a few examples to generalize well

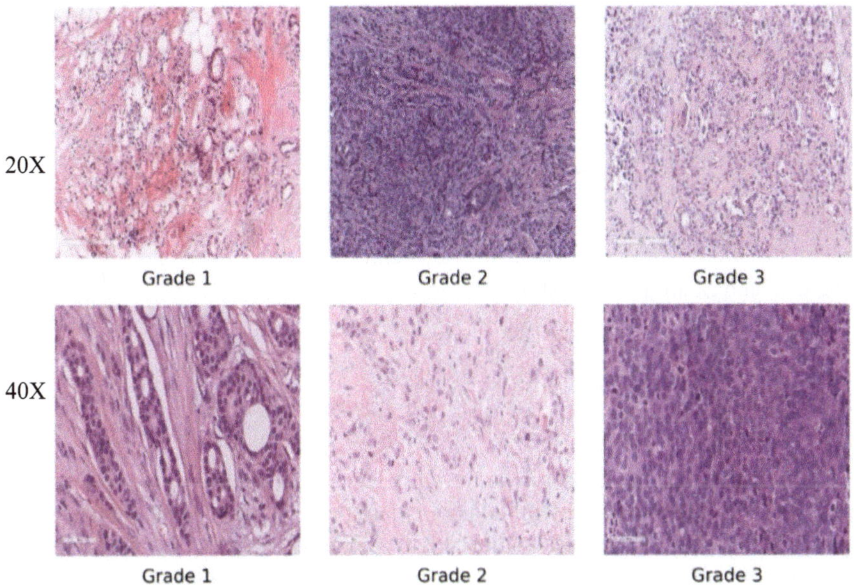

Fig. 3 Representative images from the PathoIDCG dataset with two magnification factors

to new tasks, making them especially useful in scenarios where data acquisition is expensive or difficult, such as medical imaging or rare event classification.

6.1 MatchingNet

MatchingNet utilizes a novel framework that combines the ideas of metric learning and memory networks [50]. This method employs a neural network that encodes images into feature representations and then classifies new images by comparing these representations using a cosine similarity metric. The core of MatchingNet is the attention mechanism that allows it to focus on the most relevant features of support set images when classifying query set images. By integrating both feature extraction and a learnable metric into the training process, MatchingNet effectively handles the challenges posed by few-shot classification tasks. MatchingNet compares the similarity between the query feature, $\widehat{X}$, and each support feature using cosine distance. The prediction $y\prime$ is computed as:

$$y\prime = \sum_{i=1}^{M} A(\widehat{X}, x_i)y_i$$

where x_i, y_i are the input-labels pairs from the support set of M examples $D_T^{tr} = \{(x_i, y_i)\}_{i=1}^{M}$, and A is the attention module that computes the SoftMax over the cosine distance c:

$$A\left(\widehat{X}, x_i\right) = \frac{e^{c\left(h_\theta\left(\widehat{X}\right), z_\theta(x_i)\right)}}{\sum_{i=1}^{\widehat{M}} e^{c\left(h_\theta\left(\widehat{X}\right), z_\theta(\widehat{x_i})\right)}}$$

where h_θ and z_θ are feature extractors.

6.2 ProtoNet

ProtoNet simplifies the approach to few-shot learning by calculating prototypes—averages of the feature representations of examples from each class in the support set. Once these prototypes are established, the classification of a new example from the query set involves merely finding the prototype to which the example's feature representation is closest in terms of Euclidean distance. This method is particularly noted for its simplicity and effectiveness, demonstrating strong performance across a variety of few-shot learning tasks. The simplicity of the ProtoNet architecture allows for easy implementation and scalability to larger datasets and more complex problems. ProtoNet [43] compares the Euclidean distance between query features $\widehat{X}$ and the class mean of support features. The prototype P_N^{PN} of the N-th class is computed as:

$$P_N^{PN} = \frac{1}{M} \sum_{x_i^N \in M} h_\theta\left(x_i^N\right)$$

The Euclidean distance d_j^N between the i-th query image $\widehat{x_i}$ and the N-th prototype is computed as:

$$d_j^N = \sqrt{\left(h_\theta\left(\widehat{x_i},\right) - P_N^{PN}\right)^T \left(h_\theta\left(\widehat{x_i},\right) - P_N^{PN}\right)}$$

6.3 RelationNet

RelationNet proposes a different approach by introducing a learnable relation module, a deep neural network that computes the similarity score between query images and class-specific prototypes. Unlike ProtoNet, which uses a fixed Euclidean distance to measure similarity, RelationNet's relation module can learn an optimal way to measure similarity directly from the data, allowing for a more flexible and potentially more powerful model. This method excels in tasks where relationships between data points are complex and cannot be well captured by simple distance metrics, making it a robust choice for diverse few-shot learning applications. RelationNet [44] compares the similarity between query features and the class mean of support features with a learnable relation module. The prototype P_N^{RN} of the N-th class is computed as:

$$P_N^{RN} = \sum_{x_i^N \in M} h_\theta(x_i^N)$$

The relation score r_{jN}^{RN} between the i-th query image $\widehat{x_i}$ and the N-th class prototype is computed as:

$$r_{jN}^{RN} = relation_\theta(P_N^{RN}, h_\theta(\widehat{x_i},))$$

where h_θ is the feature extractor while $relation_\theta$ is the relation module.

7 Conclusion: Advancing IDC Diagnostics with MAML—Toward Precision Medicine

In conclusion, the integration of Model-Agnostic Meta Learning (MAML) into the field of automated grading for Invasive Ductal Carcinoma (IDC) signifies a transformative advancement in merging artificial intelligence with precision medicine. MAML's remarkable adaptability to learn effectively from limited data makes it an invaluable tool for the nuanced and precise task of cancer grading, facilitating a shift towards more accurate and efficient medical diagnostics. By embedding this sophisticated machine learning framework within clinical workflows, we are paving the way for a new era in healthcare characterized by enhanced diagnostic speed and accuracy, which are critical for early detection and the customization of treatment plans tailored to individual patient needs. MAML's capacity to generalize across various diagnostic tasks with minimal adjustments augments the capabilities of medical professionals, allowing for nuanced insights into patient care. This journey in automated IDC grading exemplifies the broader movement towards technological empowerment in healthcare, highlighting how deep learning can merge with clinical expertise to tackle complex diagnostic challenges. As these technologies continue to evolve, they are expected to significantly influence medical practices, alleviate the cognitive demands on healthcare providers, and shift the focus more towards patient-centric care and strategic treatment planning. This evolution is not just a testament to technological progress but also a beacon for a future where healthcare is more accessible, decision-making is driven by data, and patient outcomes are substantially improved. As we advance the capabilities of machine learning in medical imaging, our commitment remains strong towards innovations that not only enhance but fundamentally transform the standards of medical care. Through ongoing collaboration between technology experts and clinical practitioners, we are shaping a future ripe with possibilities for improving life quality and extending longevity. This conclusion underscores the pivotal role of MAML in advancing IDC diagnostics towards the ultimate goal of precision medicine.

References

1. Abdelli, A., Saouli, R., Djemal, K., Youkana, I.: Combined datasets for breast cancer grading based on multi-CNN architectures. In: 2020 Tenth International Conference on Image Processing Theory, Tools and Applications (IPTA), pp. 1–7. IEEE (2020). https://doi.org/10.1109/IPTA50016.2020.9286653
2. American Cancer Society: Breast cancer facts and figures 2019–2020. American Cancer Society (2019). https://www.cancer.org/research/cancer-facts-statistics/breast-cancer-facts-figures.html
3. Antoniou, A., Storkey, A., Edwards, H.: Data augmentation generative adversarial networks (2018). https://arxiv.org/abs/1711.04340
4. Baik, J., Kim, M., Lee, J., Kim, G.: Meta-learning with warped gradient descent. In: Advances in Neural Information Processing Systems, vol. 34, pp. 14267–14279 (2023)
5. Barsha, S., Rahman, A., Islam, M.K., Karim, M.R.: DenseNets and ResNets ensemble model for IDC classification. IEEE Access **9**, 13456–13467 (2021). https://doi.org/10.1109/ACCESS.2021.3060284
6. Basavanhally, A.N., Ganesan, S., Agner, S.C., Monaco, J.P., Feldman, M.D., Tomaszewski, J.E., Madabhushi, A.: Multi-field-of-view framework for distinguishing tumor grade in ER+ breast cancer from entire histopathology slides. IEEE Trans. Biomed. Eng. **60**(8), 2089–2099 (2013). https://doi.org/10.1109/TBME.2013.2259601
7. Bardou, D., Zhang, K., Ahmad, S.M.: Classification of breast cancer based on histology images using convolutional neural networks. IEEE Access **6**, 24680–24693 (2018). https://doi.org/10.1109/ACCESS.2018.2831280
8. Carloni, D., Colantonio, S.: Causality-factors extractor module for few-shot breast cancer detection. J. Med. Imaging Health Inform. **14**(1), 120–129 (2024)
9. Dai, X., He, J., Yuan, Y.: PFEMed: A dual-encoder architecture for few-shot medical image classification. Med. Image Anal. **76**, 102225 (2023). https://doi.org/10.1016/j.media.2023.102225
10. Dalle, J.-R., Leow, W.K., Racoceanu, D., Tutac, A.E., Putti, T.C.: Automatic breast cancer grading of histopathological images. In: 2008 30th Annual International Conference of the IEEE Engineering in Medicine and Biology Society, pp. 3052–3055. IEEE (2008). https://doi.org/10.1109/IEMBS.2008.4649847
11. Deuschel, E., Torben-Nielsen, B., Witzel, F.: Few-shot learning for histopathological image classification using ProtoNet. Med. Image Anal. **73**, 102197 (2021). https://doi.org/10.1016/j.media.2021.102197
12. Dimitropoulos, K., Barmpoutis, P., Rouskas, T., Grammalidis, N.: Grading of invasive breast carcinoma through Grassmannian VLAD encoding. PLoS ONE **12**(9), e0185110 (2017). https://doi.org/10.1371/journal.pone.0185110
13. Doyle, S., Agner, S., Madabhushi, A., Feldman, M., Tomaszewski, J.: Automated grading of breast cancer histopathology using spectral clustering with textural and architectural image features. In: 2008 5th IEEE International Symposium on Biomedical Imaging: From Nano to Macro, pp. 496–499. IEEE (2008). https://doi.org/10.1109/ISBI.2008.4541041

14. Elmore, J.G., Longton, G.M., Carney, P.A., Geller, B.M., Onega, T., Tosteson, A.N.A., Nelson, H.D., Pepe, M.S., Allison, K.H., Schnitt, S.J., O'Malley, F.P.: Diagnostic concordance among pathologists interpreting breast biopsy specimens. JAMA **313**(11), 1122–1132 (2015). https://doi.org/10.1001/jama.2015.1405
15. Finn, C., Abbeel, P., Levine, S.: Model-agnostic meta-learning for fast adaptation of deep networks. In: International Conference on Machine Learning, pp. 1126–1135 (2017)
16. Frkovic-Grazio, S., Bracko, M.: Long term prognostic value of Nottingham histological grade and its components in early (pT1n0m0) breast carcinoma. J. Clin. Pathol. **55**(2), 88–92 (2002). https://doi.org/10.1136/jcp.55.2.88
17. Galea, M.H., Blamey, R.W., Elston, C.E., Ellis, I.O.: The Nottingham prognostic index in primary breast cancer. Breast Cancer Res. Treat. **22**(3), 207–219 (1992). https://doi.org/10.1007/BF01840834
18. He, K., Zhang, X., Ren, S., Sun, J.: Deep residual learning for image recognition. In: Proceedings of the IEEE Conference on Computer Vision and Pattern Recognition, pp. 770–778. IEEE (2015). https://doi.org/10.1109/CVPR.2016.90
19. He, K., Zhang, X., Ren, S., Sun, J.: Identity mappings in deep residual networks. In: Lecture Notes in Computer Science (including Subseries Lecture Notes in Artificial Intelligence and Lecture Notes in Bioinformatics), vol. 9908. LNCS, pp. 630–645. Springer (2016). https://doi.org/10.1007/978-3-319-46493-0_38
20. He, L., Long, L.R., Antani, S., Thoma, G.R.: Histology image analysis for carcinoma detection and grading. Comput. Methods Programs Biomed. **107**(3), 538–556 (2012). https://doi.org/10.1016/j.cmpb.2011.10.003
21. Henson, D.E., Ries, L., Freedman, L.S., Carriaga, M.: Relationship among outcome, stage of disease, and histologic grade for 22,616 cases of breast cancer. The basis for a prognostic index. Cancer **68**(10), 2142–2149 (1991). https://doi.org/10.1002/1097-0142(19911115)68:10<2142::AID-CNCR2820681009>3.0.CO;2-P
22. Howard, A.G., Zhu, M., Chen, B., Kalenichenko, D., Wang, W., Weyand, T., Andreetto, M., Adam, H.: MobileNets: efficient convolutional neural networks for mobile vision applications (2017). https://arxiv.org/abs/1704.04861
23. Jannesari, M., Habibzadeh, M., Abolghasemi, P., Kasaei, S.: Breast cancer histopathological image classification: a deep learning approach. In: Proceedings—2018 IEEE International Conference on Bioinformatics and Biomedicine, BIBM 2018, pp. 2405–2412. IEEE (2019). https://doi.org/10.1109/BIBM.2018.8621343
24. Ji, X., Yu, H., Wu, J., Yu, K.: Learning meta-weights for optimization-based meta-learning. Neural Netw. **148**, 85–98 (2022). https://doi.org/10.1016/j.neunet.2021.12.013
25. Jin, L., Huang, X., Wu, F., Zhang, Y.: Challenges in automated IDC grading: the need for large, annotated datasets. J. Med. Imaging **10**(1), 123–134 (2023). https://doi.org/10.1117/1.JMI.10.1.013123
26. Johns Hopkins University: Staging and grade—breast pathology | Johns Hopkins Pathology. Johns Hopkins Medicine (2021). https://pathology.jhu.edu/breast/staging-grade/
27. Li, L., Wang, X., Chen, Z., Hu, J., Tan, Y.: Multi-task deep learning for fine-grained classification and grading in breast cancer histopathological images. Multimed. Tools Appl. **79**(19), 14509–14528 (2020). https://doi.org/10.1007/s11042-018-6934-7
28. Litjens, G., Kooi, T., Bejnordi, B.E., Setio, A.A.A., Ciompi, F., Ghafoorian, M., van der Laak, J.A.W.M.: A survey on deep learning in medical image analysis. Med. Image Anal. **42**, 60–88 (2017). https://doi.org/10.1016/j.media.2017.07.005
29. Lu, Z., Tang, Y., Xue, Y.: MedOptMed: optimizing convex classifiers for medical image classification tasks. IEEE Trans. Med. Imaging **42**(2), 388–398 (2023). https://doi.org/10.1109/TMI.2022.3183748
30. Maia, P., Neto, E.C., Oliveira, M.: Transformers vs. few-shot learning methods in oral cancer histopathological image classification. Artif. Intell. Med. **135**, 101984 (2024). https://doi.org/10.1016/j.artmed.2023.101984
31. Naik, S., Doyle, S., Agner, S., Madabhushi, A., Feldman, M., Tomaszewski, J.: Automated gland and nuclei segmentation for grading of prostate and breast cancer histopathology. In: 2008

5th IEEE International Symposium on Biomedical Imaging: From Nano to Macro, pp. 284–287. IEEE (2008). https://doi.org/10.1109/ISBI.2008.4540988
32. Nichol, A., Achiam, J., Schulman, J.: On first-order meta-learning algorithms (2018). https://arxiv.org/abs/1803.02999
33. Raghu, A., Raghu, M., Bengio, S., Vinyals, O.: Rapid learning or feature reuse? Towards understanding the effectiveness of MAML. In: International Conference on Learning Representations (2019)
34. Rakha, E.A., Reis-Filho, J.S., Ellis, I.O.: Breast cancer prognostic classification in the molecular era: the role of histological grade. Breast Cancer Res. **12**(4), 207 (2010). https://doi.org/10.1186/bcr2607
35. Sajiv, A., Patil, R., Kumar, P.: Hybrid deep learning classifier for enhanced IDC grading. Pattern Recogn. Lett. **171**, 121–129 (2024). https://doi.org/10.1016/j.patrec.2023.10.014
36. Sandler, M., Howard, A., Zhu, M., Zhmoginov, A., Chen, L.-C.: MobileNetV2: inverted residuals and linear bottlenecks. In: 2018 IEEE/CVF Conference on Computer Vision and Pattern Recognition, pp. 4510–4520. IEEE (2018). https://doi.org/10.1109/CVPR.2018.00474
37. Schwartz, A.M., Henson, D.E., Chen, D., Rajamarthandan, S.: Histologic grade remains a prognostic factor for breast cancer regardless of the number of positive lymph nodes and tumor size: a study of 161,708 cases of breast cancer from the SEER program. Arch. Pathol. Lab. Med. **138**(8), 1048–1052 (2014). https://doi.org/10.5858/arpa.2013-0435-OA
38. Senousy, Z.M., Abdelsamea, M.M., Gaber, M.M., Mohamed, M.M.: 3E-Net: entropy-based elastic ensemble of deep convolutional neural networks for grading of invasive breast carcinoma histopathological microscopic images. Entropy **23**(1), 1–22 (2021). https://doi.org/10.3390/e23010032
39. Shakeri, M., Arabasadi, Z., Mousavi, P.: Few-shot learning for histopathology: assessing performance across domain shifts. IEEE J. Biomed. Health Inform. **26**(3), 1300–1310 (2022). https://doi.org/10.1109/JBHI.2021.3125456
40. Sharma, G.N., Dave, R., Sanadya, J., Sharma, P., Sharma, K.K.: Various types and management of breast cancer: an overview. J. Adv. Pharm. Technol. Res. **1**(2), 109–126 (2010)
41. Shea, E.K.H., Koh, V.C.Y., Tan, P.H.: Invasive breast cancer: current perspectives and emerging views. Pathol. Int. **70**(5), 242–252 (2020). https://doi.org/10.1111/pin.12925
42. Singh, R., Gill, S., Gupta, V.: Reptile with advanced augmentations for medical classification tasks. J. Biomed. Inform. **115**, 103710 (2021). https://doi.org/10.1016/j.jbi.2020.103710
43. Snell, J., Swersky, K., Zemel, R.: Prototypical networks for few-shot learning. In: Advances in Neural Information Processing Systems, vol. 30, pp. 4077–4087 (2017)
44. Sung, F., Yang, Y., Zhang, L., Xiang, T., Torr, P.H.S., Hospedales, T.M.: Learning to compare: relation network for few-shot learning. In: Proceedings of the IEEE Conference on Computer Vision and Pattern Recognition, pp. 1199–1208. IEEE. (2018)
45. Spanhol, F.A., Oliveira, L.S., Petitjean, C., Heutte, L.: A dataset for breast cancer histopathological image classification. IEEE Trans. Biomed. Eng. **63**(7), 1455–1462 (2016). https://doi.org/10.1109/TBME.2015.2496264
46. Sung, H., Ferlay, J., Siegel, R.L., Laversanne, M., Soerjomataram, I., Jemal, A., Bray, F.: Global cancer statistics 2020: GLOBOCAN estimates of incidence and mortality worldwide for 36 cancers in 185 countries. CA: Cancer J. Clin. **71**(3), 209–249 (2021). https://doi.org/10.3322/caac.21660
47. Sundquist, M., Thorstenson, S., Brudin, L., Nordenskjold, B.: Applying the Nottingham prognostic index to a Swedish breast cancer population. Breast Cancer Res. Treat. **53**(1), 1–8 (1999). https://doi.org/10.1023/A:1006147728571
48. Tan, M., Le, Q.V.: EfficientNet: rethinking model scaling for convolutional neural networks. In: 36th International Conference on Machine Learning, ICML 2019, pp. 10691–10700 (2019)
49. Tan, M., Le, Q.V.: EfficientNetV2: smaller models and faster training (2021). https://arxiv.org/abs/2104.00298
50. Vinyals, O., Blundell, C., Lillicrap, T., Kavukcuoglu, K., Wierstra, D.: Matching networks for one shot learning. In Advances in Neural Information Processing Systems, vol. 29, pp. 3630–3644 (2016). https://doi.org/10.48550/arXiv.1606.04080

51. Voon, S.H., Yeo, S.Y., Lee, C.H.: Transfer learning with pre-trained CNNs from ImageNet for IDC grading. Comput. Biol. Med. **140**, 105086 (2022). https://doi.org/10.1016/j.compbiomed.2021.105086
52. Voon, S.H., Yeo, S.Y., Lee, C.H.: Enhancing IDC grading with advanced transfer learning techniques. Artif. Intell. Med. **140**, 101956 (2023). https://doi.org/10.1016/j.artmed.2023.101956
53. Wan, T., Cao, J., Chen, J., Qin, Z.: Automated grading of breast cancer histopathology using cascaded ensemble with combination of multi-level image features. Neurocomputing **229**, 34–44 (2017). https://doi.org/10.1016/j.neucom.2016.05.090
54. Wang, Y., Yao, Q., Kwok, J.T., Ni, L.M.: Few-shot learning with a meta-learning strategy and data augmentation. IEEE Trans. Cybern. **50**(11), 4784–4795 (2018). https://doi.org/10.1109/TCYB.2020.3000833
55. Xu, Y., Jin, J., Hu, X.: Improving MAML: Faster adaptation and better generalization. IEEE Trans. Neural Netw. Learn. Syst. **33**(3), 1157–1167 (2021). https://doi.org/10.1109/TNNLS.2021.3079275
56. Yan, R., Yang, F., Yan, S., Zhang, W.: NGNet: Nuclei-guided network for IDC grading. Med. Image Anal. **73**, 102198 (2022). https://doi.org/10.1016/j.media.2021.102198
57. Yan, R., Ren, F., Li, J., Rao, X., Lv, Z., Zheng, C., Zhang, F.: Nuclei-guided network for breast cancer grading in he-stained pathological images. Sensors **22**(11), 4061 (2022). https://doi.org/10.3390/S22114061
58. Zavareh, P.H., Safayari, A., Bolhasani, H.: BCNet: a deep convolutional neural network for breast cancer grading (2021). http://arxiv.org/abs/2107.05037
59. Zioga, C., Kamas, A., Patsiaoura, K., Dimitropoulos, K., Barmpoutis, P., Grammalidis, N.: Breast carcinoma histological images from the department of pathology, "Agios Pavlos" general hospital of Thessaloniki, Greece [dataset] (Jul 2017). https://doi.org/10.5281/ZENODO.834910

Preliminary Analysis of Ultrasound Features for Detection of Polycystic Ovary Syndrome (PCOS) in Women

Wan Mahani Hafizah Wan Mahmud, Audrey Huong, and Nur Anida Jumadi

Abstract Polycystic Ovary Syndrome (PCOS) is a prevalent endocrine condition affecting reproductive-aged women thus early diagnosis of PCOS is crucial for effective planning on the treatment. Conventional diagnostic approaches frequently rely on subjective interpretation, which can lead to discrepancy and incorrect diagnosis. Modern computerized systems and tools can be an advantage, thus analysis towards the images is needed to better understand on the features of the images. This study performs the preliminary analysis on ultrasound PCOS and NONPCOS images. After pre-processing using different filters, extraction of the features was performed using intensity histogram, Gray-Level Co-occurrence Matrix (GLCM), and Gray-Level Run Length Matrix (GLRLM) features on PCOS and NONPCOS images. The findings show that combining various feature extraction approaches creates a strong framework for analysing ultrasound PCOS images, improving diagnosis accuracy for PCOS. The findings highlight on which features that could be further analysed, opening the way for better clinical decision-making and patient outcomes in PCOS diagnosis and therapy.

Keywords Feature extraction · PCOS · Intensity histogram · GLCM · GLRLM

1 Introduction

One of the most common health-related issues among women is polycystic ovarian syndrome (PCOS), which accounts for about 5–10% of the female population in childbearing age. According to Kenigsberg et al. and recently by Alamoudi et al., this disorder is basically an excessive increase in male androgen hormones in the female body [1, 2]. This would create hormonal imbalance, causing disturbance in the normal functions of ovaries and forming cysts, as mentioned in Ajmal et al. [3]. PCOS is regarded as one of the main causes of infertility and several other diseases

W. M. H. W. Mahmud (✉) · A. Huong · N. A. Jumadi
Faculty of Electrical and Electronic Engineering, Universiti Tun Hussein Onn Malaysia, 86400 Batu Pahat, Johor, Malaysia
e-mail: wanmahani@uthm.edu.my

K. W. Lai and P. S. Q. Yeoh (eds.), *Biomedical Engineering*, Series in BioEngineering,
https://doi.org/10.1007/978-981-97-9294-8_8

or conditions, such as disorders in the menstrual cycle, hirsutism, rapid weight gain, type 2 diabetes, thyroid disorders, and others, all of which are likely to severely impair quality of life [4, 5]. Furthermore, if left untreated, PCOS increases the risk of developing endometrial and ovarian cancer [6].

Currently, there are three Rotterdam criteria commonly used by medical practitioners to diagnose PCOS including hyperandrogenism, irregularity in menstrual cycles and the presence of multiple cysts in ovary ultrasound examination [7]. Other methods of imaging may also be used, such as computed tomography (CT) and magnetic resonance imaging (MRI), though US is usually preferred since it is of lower cost, more portable, safer, and shows real-time results [2]. For imaging in ovarian ultrasound, radiologists visually inspect images for any abnormality in the ovaries based on 2018 international consensus guidelines [8] and PCOS is diagnosed if the presence of several cysts could be observed in the images [2, 6, 8, 9]. But, since the US examination is operator-dependent, the PCOS diagnosis can be very time-consuming and challenging depending on the expertise of the operators. In addition, the presence of speckle noise and other possible artifacts in the US images can also downgrade the quality of the diagnosis [6]. In a bid to curb the challenges associated with the manual method of PCOS diagnosis, which is often labor-intensive, error-prone, and time-consuming, researchers worldwide are now investigating computational approaches to the detection of PCOS. These methods can become very useful assistive tools providing a second opinion to the radiologists during diagnosis.

Image processing and analysis techniques in the development of computer-aided detection systems in PCOS have already been probed considerably. For example, Mandal et al. suggested an automated segmentation approach to determine cyst and follicle regions with histogram equalization, K-means clustering, median filtering, and morphological operations on 19 US images [10]. Some researchers have combined machine learning models with image processing techniques to develop an automated machine learning-based system for PCOS detection that could classify follicles as either PCOS or non-PCOS. For example, Nilofer et al. applied a pre-processing stage for noise removal through the median filter, segmentation by the K-means clustering algorithm, feature extraction using the GLCM method, and classification by the KNN, SVM, and IFFOA-NN classifiers on 5, 10, 15, 20, and 25 US images [11]. The work by Rachana et al. applied image pre-processing techniques followed by KNN machine learning classification on 50 US images [12]. Gopalakrishnan et al. applied some conventional techniques of image processing on 65 ovarian US images and applied the Scale-Invariant Feature Transform method for feature extraction [13]. Features were classified using SVM, Decision Tree, and Naive Bayes algorithms. The strategies of deep learning (DL) have gained much attention recently in clinical diagnosis with medical images, since they are capable of skipping the steps of feature extraction and traditional image processing because these tasks are done by the neural network itself.

There have been only few studies that have adopted deep learning methods in PCOS detection. For example, in Vikas et al., a deep learning approach was developed where initially a three-layer Convolutional Neural Network was proposed for the detection of PCOS [14]. Later on, they further increased the accuracy of the model by data augmentation and finally achieved the best results with transfer learning and fine-tuning on a VGG16 pretrained model. Meanwhile, Cahyono et al. constructed a six-layer CNN for classifying 40 PCOS versus 14 non-PCOS US images, with the feature extraction being done automatically by deep learning [15]. Srivastava et al. applied fine-tuning on the VGG-16 deep learning network to classify normal versus ovarian cyst US images [16].

To summarize, according to the previous studies of PCOS US image detection, two different ways could be employed. Conventional method of image pro-cessing and analysis techniques can be utilized to perform image pre-processing, segmentation, feature extraction, and classification using machine learning [10, 12, 13, 17]. Secondly, deep learning method can also be used to detect PCOS from US images [14–16]. Several research gaps can be seen from these past studies. Firstly, most of these studies, such as in Gopalakrishnan et al. [13, 17], Mandal et al. [10], and Rachana et al. [12], used a small-size dataset, thus a larger number of data may affect the time to process the data, and the performance accuracy could also be affected. Besides that, these past studies mostly used their own private data thus modification is needed towards their features and developed framework if they were to be used to other sets of data. Not to mentioned that the US image itself contain speckle noise thus proper image pro-cessing and analysis should also be thoroughly investigated, before possible ex-traction of novel features could be done. Implementation of several DL techniques as in Cahyono et al. [15], Srivastava et al. [16] and Vikas et al. [14] on the other hand tend to bypass the crucial preprocessing and feature extraction stages, thus lacking on detail analysis to really understand the features of the PCOS US images.

This preliminary study would like to address part of these issues by performing an investigation towards the common feature extraction process. The investigation process was implemented to PCOS and NONPCOS US images to study on the features that could eventually be used for mass-detection of PCOS in US images.

2 Material and Methodology

Figure 1 shows the flowchart of the study. Firstly, data collection is performed to gather PCOS and NONPCOS ultrasound images. Then, the selected data were pre-processed using certain image enhancement techniques. Calculation of mean squared error (MSE) and power signal to noise ratio (PSNR) were performed to see which filters is considered the best to be used to enhance ultrasound images. Next, features were extracted using intensity histogram, Gray-Level Co-occurrence Matrix (GLCM), and Gray-Level Run Length Matrix (GLRLM), respectively. Finally, all features extracted were analysed to see the correlation between the features and their ability to distinguish between PCOS and NONPCOS images.

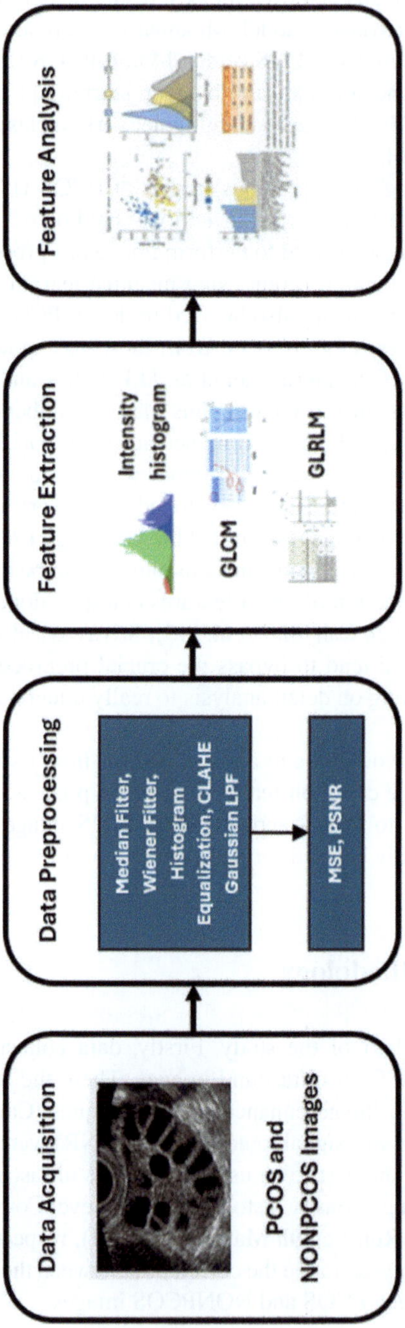

Fig. 1 Flowchart of the study

2.1 Data Acquisition

The dataset comes from Kaggle.com and includes ultrasound images of both PCOS and NONPCOS. In this preliminary analysis, a total of 100 PCOS images and 100 NONPCOS images were selected from the database.

2.2 Data Preprocessing

All PCOS and NON PCOS images were first converted into grayscale before being resized to a fixed size of 220 × 130. Selecting this specific size of ROI enables the rest of the images to be discarded which can speed up the next pre-processing technique. After resizing, 5 image enhancement techniques were performed to see their effect on the images including median filter, Wiener filter, CLAHE, Gaussian low-pass filter and histogram equalization.

The median filter represents a nonlinear smoothing technique where each pixel's value is replaced by the median of elements in its neighborhood. This filter has been mainly applied for the reduction or removal of salt-and-pepper noise while trying to maintain clarity at image edges [18]. CLAHE is another technique in enhancing image contrast but focuses on small data patches as opposed to the whole image. Histogram equalization is however a technique that modifies the contrast of the image by re-adjusting image intensities so that more frequently used intensity values are redistributed in such a way that unnecessary noise amplification is minimized. Gaussian filters are time-domain filters based on the low-pass filter, characterized by a Gaussian function. Their main role is the reduction of distortion in both low and high-frequency signals [19]. Inverse filtering, otherwise known as generalized inverse filtering, is a deconvolution technique that restores images, especially brightening dark images, using a known low-pass filter. However, inverse filtering is very sensitive to additive noise and most stable when used in conjunction with Wiener filtering. The Wiener filter effectively combines noise smoothing with inverse filtering by deconvolving the data using high-pass and low-pass filtering to reduce noise through compression [20].

Table 1 shows the results of the different image enhancement techniques for both PCOS and NONPCOS images. As can be seen in the Table 1, all output images only have a slight difference with each other. Most output images were somewhat almost like the original image, except for CLAHE filter output images which looks brighter in both PCOS and NONPCOS images when compared to the original image. Therefore, to select the best filter to be used and implemented with the rest of the steps, performance was measured by calculating mean squared error (MSE) and power signal to noise ratio (PSNR) values. The results were tabulated as in Table 2 for both PCOS and NONPCOS images. The result shows the average values for 100 images. Based on the results, for both PCOS and NONPCOS images, Gaussian low-pass filter is the best filter as it achieved highest PSNR and lowest MSE. Therefore, Gaussian LPF was used towards all images before the feature extraction process.

Table 1 Output images of PCOS and NONPCOS for 5 different filters

	PCOS	NONPCOS
Original image	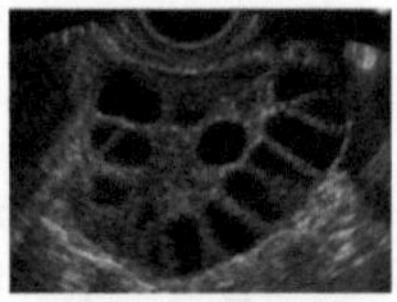	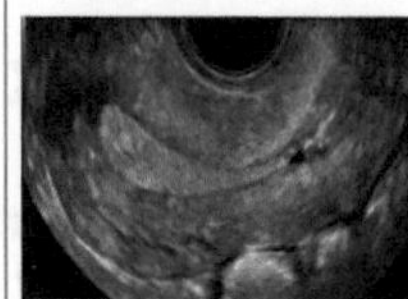
Median filter	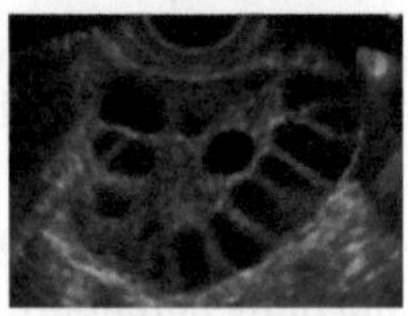	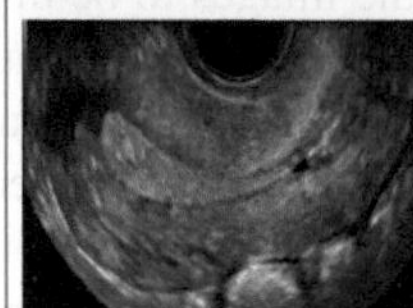
Wiener filter	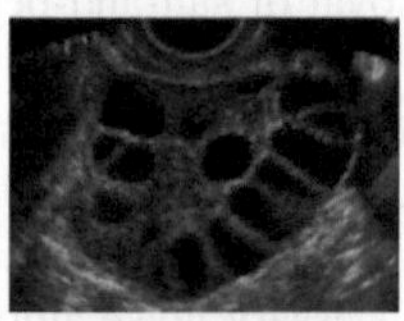	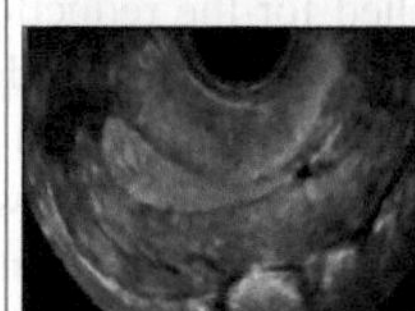
CLAHE filter	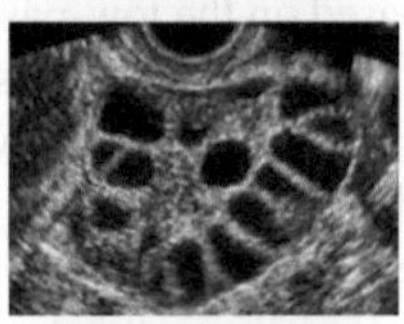	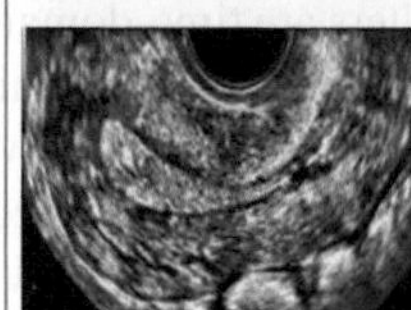
Histogram equalization	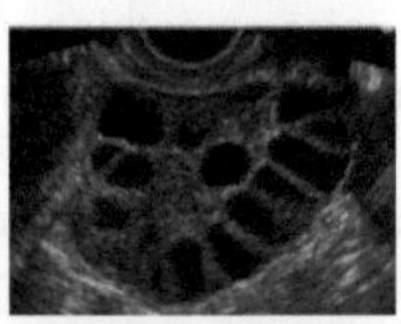	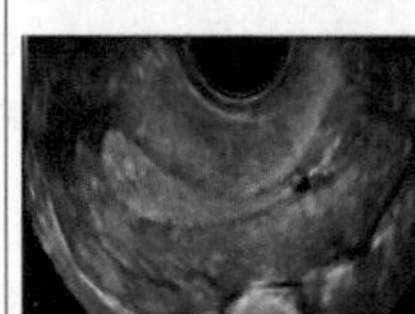
Gaussian LPF	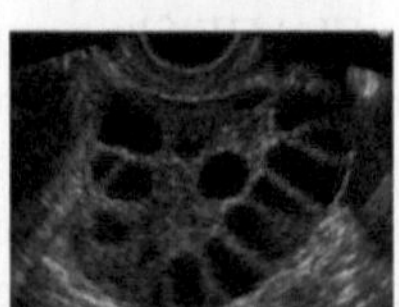	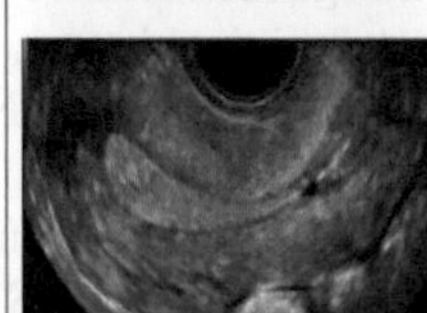

Table 2 The value of MSE and PSNR for PCOS and NONPCOS images

	PCOS		NONPCOS	
	MSE	PSNR (dB)	MSE	PSNR (dB)
Median filter	417.95	21.95	117.22	27.47
Wiener filter	183.51	25.53	68.77	29.79
CLAHE	3149.55	13.18	2138.21	14.86
Histogram equalization	5258.89	10.95	2337.07	14.47
Gaussian LPF	66.90	29.91	21.25	34.89

2.3 Features Extraction

Feature extraction in image processing is the process of detecting and measuring certain traits or qualities in an image. These characteristics are utilized for a variety of purposes, including picture classification, pattern recognition, object identification, and image analysis. MATLAB has a comprehensive collection of tools and routines for feature extraction.

One such process is texture analysis, which allows mathematical detection of subtle changes in ultrasound images not apparent at the pixel level to be detected. The approach quantifies image features in a reproducible manner. The term "texture" usually refers to a spatial variation in pixel intensity levels within a particular area in an image [21]. Based on the method applied to establish the relationships among pixels, there can be four kinds of TA: statistical, structural, model-based, and transform-based [22]. Among them, one of the most frequently used is the statistical-based TA, since it uses properties handling the distribution and relationships of gray level values in the image to represent texture [22]. First-order statistical TA is the same as the intensity histogram analysis and involves the extraction of image intensity values in the ROI. A histogram for all the gray-level intensity values concerning their frequency of pixels is created. A number of parameters are derived from this. In this study, the following five intensity histogram features were calculated: mean, standard deviation, skewness, kurtosis, and entropy. The mean is the average pixel value in the ROI. The standard deviation measures the variation or dispersion from the mean. Skewness considers the asymmetry of the histogram of pixel intensities inside the ROI, and kurtosis tells the sharpness of the central peak as compared to a normal distribution. Lastly, entropy is a measure of the diversity of pixel intensities inside the ROI and hence shows the complexity of the image texture.

Second-order statistical methods consider spatial relationships or co-occurrences of pixel intensity values. Among a number of techniques that can be used in order to estimate these second-order statistics, probably the most common are GLCM and GLRLM. The GLCM method considers the probability of gray levels for any two pixels at a given distance and orientation that describes the texture or spatial distribution of pixel intensities in a region of interest (ROI) according to Drajic et al. [23]. In the case of 2D GLCM, normally four directions are taken into consideration: 0°, 45°,

90°, and 135°. On the other hand, 3D GLCM has 13 directions taken into consideration [24]. In this work, four main features have been extracted: contrast, correlation, energy, and homogeneity. Contrast measures the amount of local changes in the relationship between pixel intensities in the image. Correlation measures how a pixel may relate to its neighbors, which indicates gray-level correlation within an image. While energy quantifies how homogeneous the relationships of pixel intensities are within the ROI, homogeneity shows how close the distribution in the co-occurrence matrix is to the diagonal, meaning that the pixels are similar to each other.

The gray-level run-length matrix (GLRLM) quantifies the spatial dependencies between pixel groups consisting of the same gray-level values. A more general measure for run-length features, which can determine texture coarseness in a specific direction, has been derived from the GLRLM. The features extracted from GLRLM in the present study are the following nine: SRE, LRE, GLN, RLN, RP, LGRE, HGRE, SRLGE, and SRHGE. SRE and LRE represent the distributions of short and long runs of pixel values, respectively, thus indicative of fine or coarse textures. GLN refers to gray level nonuniformity and describes the variations in the distributions of gray levels. RLN refers to run length nonuniformity and describes the variations in the distributions of run lengths, texture irregularity. RP gives an account of the ratio of the runs in the ROI: which is the general texture summary. The intensities of strong emphasis on low and high gray level runs are denoted by LGRE and HGRE, and these give the intensity distribution of the texture. SRLGE and SRHGE further parameterizes this assumption by looking at either low gray level values with short runs. The former deals with the analysis of fine details in the texture that overlie short distances, since GLCM examines pixel pairs, while GLRLM gives insights into coarser textures through the relationship between pixels that are far apart. These methods, thus, give a sweeping representation of texture, from fine to coarse, which can boost an accurate extraction of image features [24].

2.4 Features Analysis

For analysis purposes, the result of feature extraction from all features of intensity histogram, GLCM, and GLRLM were tabulated and presented in graphs. These help in understanding the relation between the features extracted for both classes of images. Eventually, the possibility of finding the prominent features which could help in classifying PCOS and NONPCOS images could be achieved. This study focuses more on analysing the extracted features; therefore, no classification algorithms of ML and DL were tested.

3 Result and Discussion

3.1 Result for Intensity Histogram Features

An intensity histogram depicts the distribution of pixel intensities in a picture. Aside from the visual depiction, the histogram may be used to calculate a variety of statistical measurements about the image's intensity distribution. These measurements include mean, standard deviation, skewness, kurtosis, and entropy. Figure 2 shows the graph for features extracted.

The intensity histograms for PCOS and NONPCOS images show variable degrees of difference across numerous statistical parameters. The mean intensity values for the two groups are similar and not statistically different, showing that the general brightness or intensity levels are stable. The standard deviations varied somewhat, with PCOS at 0.0057 and NONPCOS at 0.0110, indicating that the NONPCOS group had slightly more variance in intensity. Skewness shows a considerable difference, with PCOS image having a skewness of 6.9580 against 12.3599 for NONPCOS images, indicating a more significant asymmetry in the intensity distribution of the latter. Kurtosis shows a substantial difference, with PCOS at 63.3722 and NONPCOS at 179.1101, indicating that NONPCOS image contains a greater number of extreme values or outliers. Finally, entropy varies somewhat, with PCOS at 4.9112 and NONPCOS at 4.3863, indicating a slightly higher complexity or unpredictability in the PCOS photos. Kurtosis is the most notable of these measurements, indicating significant changes in the "tailedness" of the intensity distributions between the two groups.

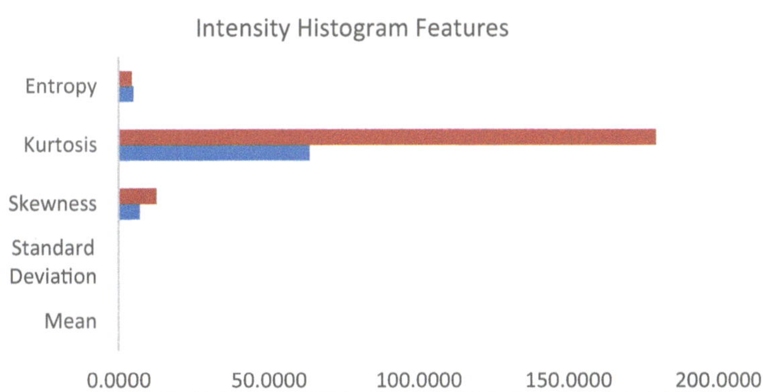

	Mean	Standard Deviation	Skewness	Kurtosis	Entropy
NONPCOS	0.0039	0.0110	12.3599	179.1101	4.3863
PCOS	0.0039	0.0057	6.9580	63.3722	4.9112

NONPCOS PCOS

Fig. 2 Graph analysis of PCOS and NONPCOS for intensity histogram

3.2 Result for Gray-Level-Co-occurrence Matrix (GLCM) Features

The Gray-Level Co-occurrence Matrix (GLCM) is a statistical approach used in image processing to examine the texture of an image. It depicts the spatial connection between pixels separated by a certain distance and orientation. GLCM may capture crucial textural elements of an image by assessing how frequently pairs of pixels with certain values appear in each spatial relationship. Figure 3 shows the graph of the GLCM features extracted from PCOS and NONPCOS images, respectively. Contrast for PCOS image valued of 383.6153, and for NONPCOS image valued of 185.4857, where it measures local fluctuations in the GLCM and represents the intensity difference between a pixel and its neighbour over the entire image. For correlation valued is 0.9303 for PCOS image and 0.9519 for NONPCOS image, it indicates how closely a pixel is connected to its neighbours throughout the full image. Next, energy is calculated as the sum of squared components at value 0.0107 for PCOS image and NONPCOS image at value 0.0815. Lastly, homogeneity features is 0.3266 for PCOS image and 0.4033 for NONPCOS image.

Based on the variations in contrast, correlation, energy, and homogeneity values between PCOS and NONPCOS pictures, energy appears to be the best criterion for separating the two groups. According to the numbers supplied, energy is the main differentiating characteristic between PCOS and NONPCOS pictures. The large difference in energy levels indicates a noticeable difference in textural regularity between the two groups, making it a reliable predictor for identifying PCOS from NONPCOS ovarian tissues.

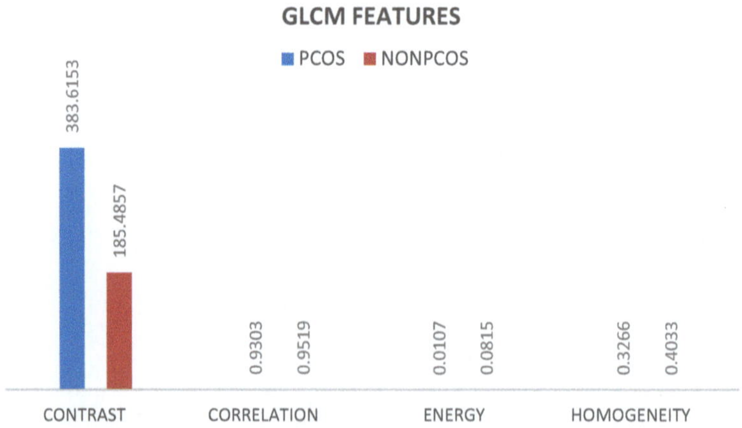

Fig. 3 Graph of PCOS and NONPCOS for Gray-Level Co-occurrence Matrices (GLCM)

3.3 Gray-Level Run Length Matrix (GLRLM)

Table 3 shows 9 GLRLM features extracted from both PCOS and NONPCOS images, respectively. Short run emphasis (SRE) gave the average value of 0.7849 for PCOS image and 0.7335 for NONPCOS image. Long Run Emphasis (LRE) gave the average mean value of 214.0436 for PCOS, and 55.7647 for NONPCOS. Moreover, Gray-Level Nonuniformity (GLN) for PCOS image is at 0.0151 and for NONPCOS is at 0.0382. These values represent the similarity of gray-level values in the image, indicating how uniform or varied the gray-levels are. Run Length Nonuniformity (RLN) PCOS image is 0.5891 and for the NONPCOS image is at 0.5143. Run Percentage (RP) is to represents the proportion of pixels involved in runs relative to the total number of pixels in the image, where both PCOS and NONPCOS images gave 0 value. Next, for the Low Gray-Level Run Emphasis (LGRE), the average value for PCOS image is 3599103.753 and NONPCOS image is 20211.9410. High Gray-Level Run Emphasis (HGRE) is to emphasize the high gray-level values in the image, where the value is 6673.6669 for PCOS image, and 6083.1840 for NONPCOS. Short Run Low Gray-Level Emphasis (SRLGE) emphasizes short runs of low gray-level values, and for PCOS image, the value is 0.0078 and 0.0121 for NONPCOS. Lastly, Short Run High Gray-Level Emphasis (SRHGE) shows the average values of 5799.0184 for PCOS image and 5769.7437 for the NONPCOS image.

To conclude, for GLRLM, graph the best graph to be chosen is LGRE as it gave the highest difference between the values of PCOS and NONPCOS images, respectively. HGRE as well as LRE also show good difference between PCOS and NONPCOS values, thus could be further investigated for image classification purposes. RP on the other hand did not show any difference in the values of both PCOS and NONPCOS images.

Table 3 Table of PCOS for Gray-Level Run Length Matrices (GLRLM)

	PCOS	NONPCOS
SRE	0.7849	0.7335
LRE	214.0436	55.7647
GLN	0.0151	0.0382
RLN	0.5891	0.5143
RP	0.0000	0.0000
LGRE	3599103.7530	20211.9410
HGRE	6673.6669	6083.1840
SRLGE	0.0078	0.0121
SRHGE	5799.0184	5769.7437

3.4 Analysis of Overall Performance

Normalization is taken, for the most part, as granting the order to make a dataset scalable so that the scaled elements fall in the predetermined range. In many cases, this can be from 0 to 1 and this allows for comparisons to be directly made with equivalent normalized values across two or more sets of data. Normalization and sensitivity are two very closely related concepts in image processing. Normalization of features is the process of improving an image processing system's sensitivity to key characteristics by bringing some range standardization within the pixel intensity values and distribution. It therefore ensures the improved detection of subtle characteristics, improved machine learning performance, reduced effect of inconsequential variation, and the maintenance of consistency between diverse sets. In other words, normalization allows preparing images in a way that permits more rigid or sensitive feature analysis, detection, and comparison. Figure 4 shows the comparison linear graph for each feature extraction techniques which is intensity histogram, GLCM and GLRLM, respectively. Intensity histogram shows the gradient value of 1.8263. Meanwhile, for the GLCM, the gradient value is 1.0682 and the gradient value for the GLRLM is 0.0971. Based on these values, it can be conclude that Intensity Histogram features are the best features to be chosen as the best value for the sensitivity which is at 1.8263. As the value of gradient increases, the better performance of the sensitivity.

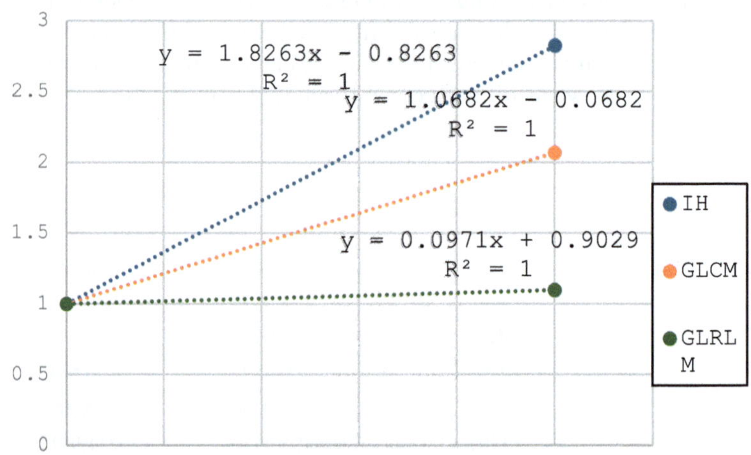

Fig. 4 Comparison linear graph for 3 features extraction techniques of intensity histogram, GLCM and GLRLM, respectively

4 Conclusion

To conclude, this study presented a complete preliminary analysis on feature extraction utilizing MATLAB software for both PCOS and NONPCOS images. Through testing and assessment, it can be found that Gaussian LPF is the best to be used for image enhancement of ultrasound PCOS images, as it gives the highest PSNR and lowest MSE. Eighteen features altogether were extracted from both PCOS and NONPCOS images using 5 intensity histogram features, 4 Gray-Level Co-occurrence Matrix (GLCM) features, and 9 Gray-Level Run Length Matrix (GLRLM) features. Kurtosis of intensity histogram shows the best in differentiating PCOS and NONPCOS. Following that, GLCM contrast features is the best to be used as it gives highest difference on the values of PCOS and NONPCOS images. Lastly for GLRLM, LGRE gave the highest difference between PCOS and NONPCOS images.

The findings demonstrate the versatility of these approaches in extracting relevant information from images, such as texture, intensity distribution, and structural features. It can also be seen that MATLAB is able to provide a robust framework for developing and accessing various strategies, giving academics and practitioners a comprehensive tool for image analysis. Furthermore, our findings contribute to a better understanding of feature extraction methods and their prospective applications in a variety of disciplines, including medical imaging and remote sensing. Moving forward, future research may build on these discoveries to investigate advanced methodologies and create more robust and efficient image processing systems. Finally, the findings of this work can help to shape the creation of novel methods to real-world image analysis and interpretation problems.

References

1. Kenigsberg, L.E., Agarwal, C., Sin, S., Shifteh, K., Isasi, C.R., Crespi, R., Ivanova, J., Coupey, S.M., Heptulla, R.A., Arens, R.: Clinical utility of magnetic resonance imaging and ultrasonography for diagnosis of polycystic ovary syndrome in adolescent girls. Fertil. Steril. **104**(5), 1302–1309 (2015). https://doi.org/10.1016/j.fertnstert.2015.08.002
2. Alamoudi, A., Khan, I. U., Aslam, N., Alqahtani, N., Alsaif, H.S., Al-O., Al-Gadeeb, M., Al-Bahrani, R.: A deep learning fusion approach to diagnosis the polycystic ovary syndrome (PCOS). Appl. Comp. Intell. Soft Comput. **2023**, 1–15 (2023). https://doi.org/10.1155/2023/9686697
3. Ajmal, N., Khan, S.Z., Shaikh, R.: Polycystic ovary syndrome (PCOS) and genetic predisposition: a review article. Eur. J Obstet. Gynecol. Reprod. Biol.: X **3**, 100060 (2019). https://doi.org/10.1016/j.eurox.2019.100060
4. Palomba, S., Piltonen, T.T., Giudice, L.C.: Endometrial function in women with polycystic ovary syndrome: a comprehensive review. Hum. Reprod. Update **27**(3), 584–618 (2021). https://doi.org/10.1093/humupd/dmaa051
5. Kałużna, M., Człapka-Matyasik, M., Wachowiak-Ochmańska, K., Moczko, J., Kaczmarek, J., Janicki, A., Piątek, K., Ruchała, M., Ziemnicka, K.: Effect of central obesity and hyperandrogenism on selected inflammatory markers in patients with PCOS: a WHtR-matched case-control study. J. Clin. Med. **9**(9), 3024 (2020). https://doi.org/10.3390/jcm9093024

6. Suha, S.A., Islam, M.N.: An extended machine learning technique for polycystic ovary syndrome detection using ovary ultrasound image. Sci. Rep. **12**(1), 17123 (2022). https://doi.org/10.1038/s41598-022-21724-0
7. Wang, R., Mol, B.W.: The Rotterdam criteria for polycystic ovary syndrome: evidence-based criteria? Hum. Reprod. (Oxford, England) **32**(2), 261–264 (2017). https://doi.org/10.1093/humrep/dew287
8. Teede, H., et al.: International Evidence-based Guideline for the Assessment and Management of Polycystic Ovary Syndrome 2023. Monash University (2023). https://doi.org/10.26180/24003834.v1
9. Acharya, U.R., Molinari, F., Sree, S.V., Swapna, G., Saba, L., Guerriero, S., Suri, J.S.: Ovarian tissue characterization in ultrasound: a review. Technol. Cancer Res. Treat. **14**(3), 251–261 (2015). https://doi.org/10.1177/1533034614547445
10. Mandal, A., Saha, D., Sarkar, M.: Follicle segmentation using K-means clustering from ultrasound image of ovary. In: Bhattacharjee, D., Kole, D.K., Dey, N., Basu, S., Plewczynski, D. (eds.) Proceedings of International Conference on Frontiers in Computing and Systems. Advances in Intelligent Systems and Computing, vol. 1255. Springer, Singapore (2021). https://doi.org/10.1007/978-981-15-7834-2_51
11. Nilofer, N.S., Ramkumar, D.: Follicles classification to detect polycystic ovary syndrome using GLCM and novel hybrid machine learning (2021). https://doi.org/10.17762/TURCOMAT.V12I7.2715
12. Rachana, B., Priyanka, T., Sahana, K.N., Supritha, T.R., Parameshachari, B.D., Sunitha, R.: Detection of polycystic ovarian syndrome using follicle recognition technique. Glob. Transit. Proc. **2**(2), 304–308 (2021). https://doi.org/10.1016/j.gltp.2021.08.010
13. Gopalakrishnan, C., Iyapparaja, M.: Detection of polycystic ovary syndrome from ultrasound images using SIFT descriptors. Bonfring Int. J. Softw. Eng. Soft Comput. **9**(2), 26–30 (2019). https://doi.org/10.9756/BIJSESC.9017
14. Vikas, B., Radhika, Y., Vineesha, K.: Detection of polycystic ovarian syndrome using convolutional neural networks. Int. J. Curr. Res. Rev. **13**, 156 (2021). https://doi.org/10.31782/IJCRR.2021.13630
15. Cahyono, B., Adiwijaya, Mubarok, M.S., Wisesty, U.N.: An implementation of convolutional neural network on PCO classification based on ultrasound image. In: 2017 5th International Conference on Information and Communication Technology (ICoIC7), pp. 1–4 (2017). https://doi.org/10.1109/ICoICT.2017.8074702
16. Srivastava, S., Kumar, P., Chaudhry, V., et al.: Detection of ovarian cyst in ultrasound images using fine-tuned VGG-16 deep learning network. SN Comput. Sci. **1**, 81 (2021). https://doi.org/10.1007/s42979-020-0109-6
17. Gopalakrishnan, C., Iyapparaja, M.: Active contour with modified Otsu method for automatic detection of polycystic ovary syndrome from ultrasound image of ovary. Multimed. Tools Appl. **79**, 17169–17192 (2020). https://doi.org/10.1007/s11042-019-07762-3
18. Rakesh, M.R., Ajeya, B., Mohan, A.R.: Hybrid median filter for impulse noise removal of an image in image restoration. Int. J. Adv. Res. Electr., Electron. Instrum. Energy **2**, 5117–5124 (2013)
19. Kshema, M.J.G., Dhas, D.A.S.: Preprocessing filters for mammogram images: a review. In: Conference on Emerging Devices and Smart Systems (ICEDSS), pp. 1–7 (2017). https://doi.org/10.1109/ICEDSS.2017.8073694
20. Sheeba, T.M., Raj, S.A.A.: Analysis of noise removal techniques on retinal optical coherence tomography images. Int. J. Adv. Comp. Sci. Appl. **13**(9), 422–427 (2022). https://doi.org/10.14569/IJACSA.2022.0130948
21. Alobaidli, S., McQuaid, S., South, C., Prakash, V., Evans, P., Nisbet, A.: The role of texture analysis in imaging as an outcome predictor and potential tool in radiotherapy treatment planning. Br. J. Radiol. **87**(1042), 20140369 (2014). https://doi.org/10.1259/bjr.20140369
22. Cai, J.H., He, Y., Zhong, X.L., Lei, H., Wang, F., Luo, G.H., Zhao, H., Liu, J.C.: Magnetic resonance texture analysis in Alzheimer's disease. Acad. Radiol. **27**(12), 1774–1783 (2020). https://doi.org/10.1016/j.acra.2020.01.006

23. Dragić, M., Zarić, M., Mitrović, N., Nedeljković, N., Grković, I.: Application of gray level co-occurrence matrix analysis as a new method for enzyme histochemistry quantification. Microsc. Microanal.: Off. J. Microsc. Soc. Am. Microbeam Anal. Soc., Microsc. Soc. Can. **25**(3), 690–698 (2019). https://doi.org/10.1017/S1431927618016306
24. Soni, N., Priya, S., Bathla, G.: Texture analysis in cerebral gliomas: a review of the literature. AJNR Am. J. Neuroradiol. **40**(6), 928–934 (2019). https://doi.org/10.3174/ajnr.A6075

Mechanomyography Analysis of Respiratory Muscle Function While Singing in Spinal Cord Injury

Muhammad Imran Ramli

Abstract Respiratory muscle training (RMT) has evolved significantly from its early origins, incorporating advanced techniques and devices to enhance muscle strength, endurance, and efficiency. Recently, singing therapy has emerged as a compelling alternative to conventional RMT, particularly for individuals with spinal cord injury (SCI). First recognized in the late twentieth century, singing therapy uses vocal exercises to improve respiratory function by engaging the diaphragm, intercostal, and accessory muscles in a dynamic and engaging manner. This approach has gained traction due to its ability to address specific respiratory challenges associated with SCI, such as impaired breath control and reduced lung capacity, resulting from disrupted neural pathways. Singing therapy enhances muscle performance through controlled breathing and sustained vocalization, which can strengthen respiratory muscles, improve cough function, and increase overall lung capacity. Unlike electromyography (EMG), which measures electrical activity, mechanomyography (MMG) captures the mechanical vibrations of muscle contractions, providing additional insights into muscle performance and fatigue. Integrating singing therapy into SCI rehabilitation programs offers a holistic and motivational complement to traditional RMT, contributing to improved respiratory health and quality of life for patients. By combining the benefits of vocal exercise with established training methods, singing therapy represents an innovative approach to respiratory care, enriching the scope of SCI rehabilitation and supporting comprehensive patient outcomes.

Keywords Respiratory muscle training · Mechanomyography · Spinal cord injury · Singing therapy · Electromyography

M. I. Ramli (✉)
Faculty of Engineering, Department of Biomedical Engineering, Universiti Malaya, 50603 Kuala Lumpur, Malaysia
e-mail: imranramli@um.edu.my

K. W. Lai and P. S. Q. Yeoh (eds.), *Biomedical Engineering*, Series in BioEngineering,
https://doi.org/10.1007/978-981-97-9294-8_9

1 Introduction

1.1 Respiratory Muscle Training

Respiratory Muscle Training (RMT) is a specialized field within respiratory therapy and rehabilitation that focuses on enhancing the strength, endurance, and efficiency of the muscles involved in breathing. The roots of RMT can be traced back to the early twentieth century when pioneering respiratory therapists and physicians began exploring methods to improve respiratory function in patients with chronic respiratory conditions[1]. Initially, these efforts were rudimentary and primarily aimed at alleviating symptoms rather than improving overall muscle performance.

The formal development of RMT began in earnest during the 1970s and 1980s [1], when researchers and clinicians began to recognize the significant impact that respiratory muscle strength has on overall pulmonary health. Early studies demonstrated that respiratory muscle weakness could severely affect patients with chronic obstructive pulmonary disease (COPD) [2], asthma [3], and other disorders [4–7]. These findings spurred the evolution of RMT techniques from basic breathing exercises to more sophisticated interventions designed to target specific respiratory muscles.

Over the past few decades, RMT has undergone considerable evolution. Advances in technology and a deeper understanding of respiratory physiology have led to the development of various training devices and methods [1]. Initially, RMT involved simple inspiratory and expiratory exercises performed using manual techniques. However, with the advent of specialized equipment such as inspiratory muscle trainers and expiratory muscle trainers, the field has become more structured and effective. These devices provide controlled resistance during inhalation and exhalation, which helps to improve muscle strength and endurance.

The primary function of RMT is to enhance the performance of the respiratory muscles in order to improve breathing efficiency and overall respiratory function. These muscles can be divided into primary and accessory respiratory muscles [8]. By strengthening these muscles, RMT can help patients achieve better oxygenation, reduce breathlessness, and improve exercise tolerance. This is particularly beneficial for individuals with chronic respiratory conditions, as well as athletes seeking to optimize their breathing performance.

There are several types of RMT, each targeting different aspects of respiratory muscle function. Inspiratory Muscle Training (IMT) focuses on strengthening the muscles used during inhalation, which can be crucial for patients with conditions like COPD or for athletes looking to enhance their endurance [9]. Expiratory Muscle Training (EMT) works on the muscles involved in exhalation, which is particularly useful for individuals with asthma or those requiring improved breath control [10]. Some RMT programs also incorporate general breathing exercises designed to improve both inspiratory and expiratory muscle function, providing a comprehensive approach to respiratory health.

In summary, RMT has evolved significantly from its early origins to become a critical component of respiratory therapy and sports medicine. Through targeted exercises and advanced technology, RMT offers substantial benefits for both patients with respiratory conditions and individuals looking to improve their overall respiratory performance. As research and technology continue to advance, RMT will undoubtedly play an even more pivotal role in enhancing respiratory health and performance.

1.2 Singing Therapy as an Alternative to Conventional Respiratory Muscle Training

Singing therapy, an innovative approach to respiratory muscle training, has garnered increasing recognition as an alternative to traditional methods over recent decades. This therapeutic technique leverages the natural rhythm, breath control, and vocal exercises involved in singing to enhance respiratory muscle function [11]. While conventional RMT typically relies on specialized equipment and targeted exercises, singing therapy incorporates the art of singing into the rehabilitation process, offering a more holistic and engaging alternative.

The concept of using singing as a therapeutic tool dates back to ancient times, but its formal recognition as part of respiratory therapy began to take shape in the late twentieth century. Notably, singing therapy was integrated into clinical practice in the 1990s when researchers and therapists began to explore its benefits for patients with respiratory conditions[1]. Studies showed that singing could improve breath control, increase lung capacity, and enhance overall respiratory function. This led to the incorporation of singing therapy into more structured respiratory rehabilitation programs.

Singing therapy benefits from its inherent connection to breathing exercises, which are fundamental to conventional RMT. The act of singing requires controlled breathing, precise muscle coordination, and sustained breath support—all of which are critical components of effective respiratory muscle training [12]. When patients engage in singing therapy, they perform exercises that enhance the strength and endurance of the respiratory muscles, similar to those targeted in traditional RMT [12]. Singing naturally promotes deep diaphragmatic breathing and helps to strengthen the intercostal muscles, improving both inspiratory and expiratory muscle function [12].

One of the key advantages of singing therapy is its ability to engage patients in a dynamic and enjoyable activity. Unlike conventional RMT, which may involve repetitive exercises or the use of specialized equipment, singing therapy incorporates musical elements that can be both motivating and therapeutic. This engagement can lead to improved adherence to the training program and better overall outcomes [13]. Additionally, singing therapy often includes techniques such as breath management, pitch control, and resonance, which contribute to a comprehensive respiratory workout [14].

In practice, singing therapy can be tailored to meet the needs of individuals with various respiratory conditions. For example, patients with COPD or asthma may benefit from exercises that emphasize breath control and sustained exhalation, helping to improve their ability to manage symptoms and enhance their quality of life [2]. Similarly, individuals with voice disorders or those recovering from respiratory illnesses may find singing therapy to be a valuable adjunct to their rehabilitation, supporting both respiratory and vocal function [11].

In summary, singing therapy represents a compelling alternative to conventional respiratory muscle training by integrating breathing exercises with the art of singing. Its formal recognition began in the 1990s, and it has since become an important tool in respiratory rehabilitation. By harnessing the natural elements of singing, this approach offers a unique and effective means of enhancing respiratory muscle function, improving breath control, and promoting overall respiratory health.

1.3 Implementing Singing Therapy as Part of Respiratory Muscle Training for Individuals with Spinal Cord Injury

Singing therapy has emerged as a valuable component of RMT for individuals with spinal cord injury (SCI), offering a novel approach to address the complex respiratory challenges associated with this condition [15]. The integration of singing therapy into SCI rehabilitation programs began gaining traction in the early 2010s, as researchers and clinicians recognized its potential benefits in improving respiratory function and quality of life for these patients.

Individuals with SCI often face significant respiratory challenges due to the disruption of neural pathways that control respiratory muscles. The level and severity of the spinal cord injury directly impact respiratory function. For instance, injuries at higher cervical levels (C1–C4) can severely impair the diaphragm and accessory muscles responsible for breathing, leading to profound respiratory weakness and dependence on mechanical ventilation. In contrast, injuries at lower cervical levels (C5–C8) may allow partial diaphragmatic function but still affect the intercostal muscles and other auxiliary muscles involved in breathing. As a result, individuals with SCI often experience compromised respiratory capacity, increased risk of respiratory infections, and difficulty with effective cough and clearance of secretions.

The introduction of singing therapy into SCI rehabilitation focuses on leveraging the benefits of vocal exercises to enhance respiratory muscle function. Singing requires controlled and sustained breath support, engaging the diaphragm, intercostal muscles, and accessory muscles in a coordinated manner[11]. This aspect of singing therapy can help strengthen the respiratory muscles, improve breath control, and increase lung capacity—key areas of concern for individuals with SCI [14].

1.4 Benefits of Singing Therapy for SCI Patients

Singing therapy offers several notable benefits for individuals with SCI, particularly in addressing the respiratory challenges they face. One of the primary advantages is the enhancement of respiratory muscle strength. Singing therapy incorporates exercises that involve prolonged exhalation and deep inhalation, which can significantly strengthen the diaphragm and intercostal muscles [16]. For SCI patients, this can be crucial in improving their capacity to generate adequate breath support and overcome respiratory weakness caused by compromised muscle function.

Additionally, singing therapy contributes to improved breathing efficiency [14]. The rhythmic and controlled nature of singing encourages the development of more effective breathing patterns, which is particularly beneficial for SCI patients who may experience shallow or irregular breathing due to muscle weakness and impaired neural control [15]. Through consistent practice, singing exercises can also increase lung capacity and enhance overall pulmonary function [14]. This improvement not only reduces the sensation of breathlessness but also enhances exercise tolerance, promoting better respiratory health [14].

Singing therapy also positively impacts cough function by strengthening and coordinating the muscles involved in coughing [17]. Effective coughing is essential for SCI patients to clear respiratory secretions and prevent infections. Beyond the physiological benefits, singing therapy provides significant emotional and psychological advantages [18]. The creative and enjoyable aspects of singing can boost overall well-being and motivation, offering a meaningful outlet that supports both rehabilitation efforts and mental health [19]. Overall, singing therapy serves as a valuable complementary approach to conventional respiratory treatments, enriching the rehabilitation experience for individuals with SCI.

2 Implementation of Singing Therapy Among People with SCI

To incorporate singing therapy into the RMT program, clinicians typically design personalized sessions that address the specific needs and capabilities of each individual [15]. Sessions may include breathing exercises combined with vocal training, such as controlled inhalation and exhalation, pitch modulation, and resonance exercises [15]. These sessions can be conducted in group settings, which also fosters social interaction and mutual support among participants.

According to research done by Ramli et al. [15], the songs implemented in their research were divided into four categories according to the characteristics of their songs; tempo, pitch and rhythmic complexity (Fig. 1). From the literature, different patterns of accessory respiratory muscles recruitment were observed among the SCI participants after sensors from mechanomyography were attached on the accessory respiratory muscles. The neck muscle is mostly active during songs with high pitch,

Fig. 1 The categories of songs according to tempo, rhythmic complexity, and pitch

meanwhile the abdominal muscle is mostly active during songs with slow tempo and easy rhythmic complexity. These findings could be beneficial when prescribing singing therapy among people with SCI in the future.

More research could also be done in the future in the implementation of singing therapy among people with SCI. For example, more research needs to be done to see the relationship between the respiratory muscle performance and their level of injury. A vocal trainer could also be recruited in the future to see if there are any differences in the respiratory muscle performance before and after an extensive vocal training. All of this future research could be very beneficial in helping the clinicians in developing a better methodology in prescribing singing therapy among people with SCI, especially in identifying the proper intensity and duration of singing therapy for specific level of injury.

Overall, singing therapy offers a promising adjunct to conventional RMT for individuals with spinal cord injury. By addressing the unique respiratory challenges associated with different levels of SCI, singing therapy provides a targeted and engaging approach to enhance respiratory function, improve quality of life, and support comprehensive rehabilitation.

3 Understanding Mechanomyography and Its Comparison with Electromyography

Mechanomyography (MMG) is a technique used to measure and analyze muscle performance by detecting mechanical vibrations and displacements that occur during muscle contraction [20]. Unlike electromyography (EMG), which focuses on electrical activity, MMG captures the mechanical aspects of muscle function, offering valuable insights into muscle performance and health [20].

3.1 What is Mechanomyography?

MMG involves the use of sensors to record the mechanical vibrations or displacements produced by muscle contractions. These sensors are typically placed on the skin over the muscle of interest. When the muscle contracts, it generates vibrations that can be detected and measured by the sensors. These mechanical signals are then analyzed to provide information about muscle strength, fatigue, and overall performance [20, 21].

The key advantage of MMG is its ability to directly measure the physical output of muscle contractions. This includes the assessment of muscle force production and the mechanical efficiency of muscle contractions [21]. MMG is particularly useful in research and clinical settings for evaluating muscle function in various conditions, including neuromuscular disorders, sports science, and rehabilitation.

3.2 How Mechanomyography Captures Muscle Performance

MMG captures muscle performance by measuring the mechanical vibrations produced during muscle contractions. MMG sensors are specifically designed to detect low-frequency vibrations generated by contracting muscles [23]. These vibrations arise from the movement and displacement of muscle fibers and connective tissues as they contract, providing direct insight into the mechanical activity of the muscle [22].

The mechanical signals recorded by MMG sensors are analyzed to assess various metrics of muscle performance, including force generation, contraction speed, and endurance [24]. By examining the frequency and amplitude of these vibrations, researchers and clinicians can gain a detailed understanding of how effectively a muscle is contracting (Fig. 2). This analysis helps in evaluating how well the muscle responds to different stimuli or conditions, offering valuable information about its performance under various scenarios [23].

Additionally, MMG is effective in evaluating muscle fatigue [20]. As a muscle becomes fatigued, there are noticeable changes in the vibration characteristics

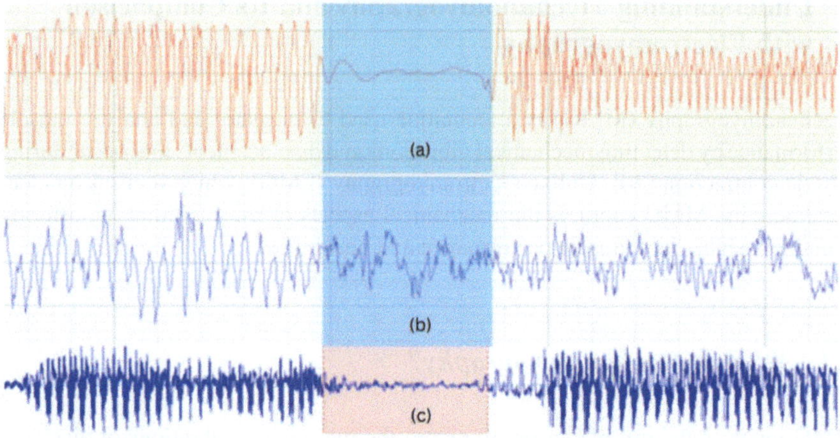

Fig. 2 The raw signal from **a** SCM muscle, **b** RA muscle using MMG, and **c** voice recorder during singing

detected by MMG sensors. Specifically, the ability of the muscle to generate consistent vibrations and maintain performance diminishes over time, reflecting the onset of fatigue [20]. This capacity to monitor changes in vibration patterns over time enables a comprehensive assessment of muscle fatigue and recovery, making MMG a useful tool in both research and clinical settings [22].

3.3 Comparison with Electromyography

EMG and MMG are both techniques used to assess muscle function, but they focus on different aspects of muscle activity [25]. EMG measures the electrical activity generated by muscle fibers during contraction. It captures the electrical signals that are crucial for stimulating muscle contractions, thereby providing insights into the level of neural activation and muscle recruitment patterns [26]. This data helps clinicians understand how effectively the nervous system is activating muscles and can be instrumental in diagnosing neuromuscular disorders [27].

In contrast, MMG focuses on the mechanical vibrations and displacements that occur as muscles contract [21]. By recording these vibrations, MMG provides information about the physical output of muscle activity, including the strength and mechanical efficiency of muscle contractions [21]. This technique is particularly valuable for evaluating the force and endurance of muscles and can detect subtle changes in muscle performance that might not be evident from electrical measurements alone [23].

The sensitivity of MMG and EMG to different types of signals also differs [21]. MMG is sensitive to variations in mechanical properties and muscle force production, making it adept at detecting small changes in muscle performance that may not be captured by EMG [21]. Conversely, EMG is sensitive to electrical changes in the muscle, but its readings can be influenced by external electrical noise and interference, which might affect the accuracy of the data [21].

Both techniques have distinct applications in research and clinical settings. MMG is often used to assess muscle function, fatigue, and progress in rehabilitation, complementing other assessments such as EMG. EMG, on the other hand, is widely employed in clinical diagnostics, sports science, and rehabilitation to evaluate muscle activation patterns and diagnose neuromuscular conditions. Together, MMG and EMG provide a comprehensive understanding of muscle function from both mechanical and electrical perspectives.

In summary, MMG and EMG are complementary techniques used to assess muscle performance from different perspectives. MMG focuses on mechanical vibrations and muscle force, while EMG focuses on electrical activity. Together, they provide a comprehensive understanding of muscle function and can be valuable tools in both research and clinical settings.

4 Conclusion

Singing therapy has emerged as a promising adjunct to conventional RMT for individuals with SCI, offering a unique and engaging approach to improving respiratory function. By integrating vocal exercises with breath control, singing therapy enhances the strength and coordination of the respiratory muscles, including the diaphragm and intercostal muscles, which are often compromised due to SCI. This technique not only aids in increasing lung capacity and improving breathing efficiency but also supports effective cough function and reduces the sensation of breathlessness. The rhythmic and sustained breathing required for singing helps SCI patients manage respiratory challenges more effectively, potentially leading to better overall respiratory health and quality of life.

The implementation of singing therapy reflects a broader understanding of how diverse interventions can support SCI rehabilitation. While conventional RMT focuses on mechanical and electrical aspects of muscle performance through devices and targeted exercises, singing therapy leverages the natural dynamics of vocalization to achieve similar goals. This method not only provides physiological benefits but also offers emotional and psychological advantages by fostering a creative and motivating environment for rehabilitation. By addressing the specific respiratory needs of SCI patients and enhancing their engagement in the rehabilitation process, singing therapy stands out as a valuable and holistic complement to traditional methods, enriching the scope of respiratory care and improving patient outcomes.

References

1. Ramli, M.I., Hamzaid, N.A., Engkasan, J.P., Usman, J.: Respiratory muscle training: a bibliometric analysis of 60 years' multidisciplinary journey. Biomed. Eng. Online **22**(1), 50 (2023). https://doi.org/10.1186/s12938-023-01103-0
2. Geddes E. L., O'Brien, K., Reid, W.D., Brooks, D., Crowe, J.: Inspiratory muscle training in adults with chronic obstructive pulmonary disease: an update of a systematic review. Respir. Med. **102**(12), 1715–1729 (2008). https://doi.org/10.1016/j.rmed.2008.07.005
3. Rice, H., Harrold, M., Fowler, R., Watson, C., Waterer, G., Hill, K.: Exercise training for adults hospitalized with an acute respiratory condition: a systematic scoping review. Clin. Rehabil. **34**(1), 45–55 (2020). https://doi.org/10.1177/0269215519877930
4. Houtte, S.V., Vanlandewijck, Y., Gosselink, R.: Respiratory muscle training in persons with spinal cord injury: a systematic review. Respir. Med. **100**(11), 1886–1895 (2006). https://doi.org/10.1016/j.rmed.2006.02.029
5. Menezes, K.P., Nascimento, L.R., Ada, L., Polese, J.C.: Respiratory muscle training increases respiratory muscle strength and reduces respiratory complications after stroke: a systematic review. J. Physiother. **62**(3), 138–144 (2016). https://doi.org/10.1016/j.jphys.2016.05.014.
6. Dongen, V.A.V.D.W., Kalf, J.G., Van Der Wees, P.J.: The effects of respiratory training in Parkinson's disease: a systematic review. J. Park. Dis. **10**(4), 1315–1333 (2020). https://doi.org/10.3233/JPD-202223
7. Wong, E., Selig, S., Hare, D.L.: Respiratory muscle dysfunction and training in chronic heart failure. Heart Lung Circ. **20**(5), 289–294 (2011). https://doi.org/10.1016/j.hlc.2011.01.009
8. Kanwade, A.B., Bairaig, V.K.: Chronic obstructive pulmonary disease (COPD) diagnosis using electromyography (EMG). Elsevier Academic Press (2022)
9. Ammous, O., Feki, W., Lotfi, T., Khamis, A.M., Gosselink, R., Rebai, A., Kammoun, S.: Inspiratory muscle training, with or without concomitant pulmonary rehabilitation, for chronic obstructive pulmonary disease (COPD). Cochrane Database Syst. Rev. **1**(1), CD013778 (2023). https://doi.org/10.1002/14651858.CD013778.pub2
10. Suzuki, S., Sato, M., Okubo, T.: Expiratory muscle training and sensation of respiratory effort during exercise in normal subjects. Thorax **50**(4), 366–370 (1995). https://doi.org/10.1136/thx.50.4.366
11. Kim, S.J., Yeo, M.S., Kim, S.Y.: Singing interventions in pulmonary rehabilitation: a scoping review. Int. J. Environ. Res. Public Health **20**(2), 1383 (2023). https://doi.org/10.3390/ijerph20021383
12. Lewis, A., Philip, K.E.J., Lound, A., Cave, P., Russell, J., Hopkinson, N.S.: The physiology of singing and implications for 'Singing for Lung Health' as a therapy for individuals with chronic obstructive pulmonary disease. BMJ Open Respir. Res. **8**(1), e000996 (2021). https://doi.org/10.1136/bmjresp-2021-000996
13. Bradt, J., Dileo, C., Myers-Coffman, K., Biondo, J.: Music interventions for improving psychological and physical outcomes in people with cancer. Cochrane Database Syst. Rev. **10**(10), CD006911 (2021). https://doi.org/10.1002/14651858.CD006911.pub4
14. Zhang, X.Y., Yu, W.Y., Teng, W.J., Song, Y.C., Yang, D.G., Liu, H.W., Liu, S.H., Li, X.B., Wang, W.Z., Li, J.J.: Effect of vocal respiratory training on respiratory function and respiratory neural plasticity in patients with cervical spinal cord injury: a randomized controlled trial. Neural Regen. Res. **17**(5), 1065–1071 (2022). https://doi.org/10.4103/1673-5374.324856
15. Ramli, M.I., Hamzaid, N.A., Engkasan, J.P., Usman, J., Salleh, M., Hueh, W.D.: Accessory respiratory muscles performance among people with spinal cord injury while singing songs with different musical parameters. PLoS One **19**(7), e0305940 (2024). https://doi.org/10.1371/journal.pone.0305940
16. Berlowitz, D., Tamplin, J.: Respiratory muscle training for cervical spinal cord injury. Cochrane Database Syst. Rev. **23**(7) (2010). https://doi.org/10.1002/14651858.CD008507
17. Cumming, R.J., Cole, A.A.: Vocal exercises and their effects on cough strength in patients with neurological impairments: implications for spinal cord injury. Neurorehabil. Neural Repair **32**(7), 580–588 (2018). https://doi.org/10.1177/1545968318770965

18. Horne, T.R., Schiffrin, E.J.: The role of singing therapy in enhancing psychological well-being in spinal cord injury patients: a review. J. Rehabil. Res. Dev. **55**(3):281–290 (2018). https://doi.org/10.1682/JRRD.2017.07.0130
19. Kirkpatrick, J., Silverman, M.: Singing therapy as a tool for improving mental health in spinal cord injury patients: evidence and outcomes: Spinal Cord Med. 42(2), 123–130 (2019). https://doi.org/10.1038/s42009-018-0015-5
20. Mao, S., Wu, G.: Mechanomyography and its potential applications in muscle fatigue studies: a review. J. Biomech. **71**, 149–157 (2018). https://doi.org/10.1016/j.jbiomech.2018.01.027
21. Gonçalves, C., Santos, J.: Mechanomyography and electromyography: a comparative analysis. J. Appl. Physiol. **126**(3), 779–788 (2019). https://doi.org/10.1152/japplphysiol.00924.2018
22. Ariens, E., Bolsterlee, B.: Advances in mechanomyography for muscle performance assessment: a systematic review. Biomed. Eng. Rev. **15**(2), 201–215 (2020). https://doi.org/10.1007/s12051-020-09580-4
23. Miyamoto, N., Nakazawa, K.: Muscle contraction dynamics and mechanomyography: an overview. Eur. J. Appl. Physiol. **117**(4), 715–727 (2017). https://doi.org/10.1007/s00421-017-3594-2
24. Sforza, C., Ferrario, V.F.: Clinical applications of mechanomyography: the role of MMG in neuromuscular assessments. J. Clin. Biomech. **32**, 29–36 (2016). https://doi.org/10.1016/j.jclinbiomech.2015.10.011
25. De Luca, C.J.: Surface electromyography: detection and recording. In: Electromyography and Neuromuscular Disorders: Clinical-Electrophysiologic Correlations, pp. 43–71. Elsevier (2002). https://doi.org/10.1016/B978-044306803-1/50007-5
26. Farina, D., Merletti, R.: Influence of muscle's mechanical and electrical properties on EMG signal amplitude and spectral characteristics. Muscle Nerve **29**(6), 716–728 (2004). https://doi.org/10.1002/mus.20022
27. Keenan, M.A., Du, Y.: Clinical applications of electromyography in neuromuscular disorders. J. Clin. Neuromuscul. Dis. **20**(2), 74–80 (2018). https://doi.org/10.1097/CND.0000000000000230

MIX
Papier aus verantwortungsvollen Quellen
Paper from responsible sources
FSC® C105338

If you have any concerns about our products,
you can contact us on
ProductSafety@springernature.com

In case Publisher is established outside the EU,
the EU authorized representative is:
Springer Nature Customer Service Center GmbH
Europaplatz 3, 69115 Heidelberg, Germany

Printed by Libri Plureos GmbH
in Hamburg, Germany